Women Changing Science

Voices from a Field in Transition

Mary Morse

PERSEUS PUBLISHING

Cambridge, Massachusetts

Perseus Publishing books at special discounts for bulk
purchases in the U.S. by corporations, instututions, and other orginizations.
For more information, please contact the Special Markets Department at the
Perseus Books Group, 11 Cambridge, MA 02142, or call (800) 255-1514 or
(617) 252-5298, or e-mail J.mccrary@perseusbooks.com

ISBN 0-7382-0615-6

© 1995 Mary Morse

Published by Perseus Publishing
A Member of the Perseus Books Group

10 9 8 7 6 5 4 3 2 1

Printed in the United States of America

Women Changing Science

Voices from a Field in Transition

Preface

The original impetus for this book came from an idea proposed to me by Lynette Lamb, one of my editors at *Utne Reader* magazine. For several years, she was keeping tabs on issues of women and science, and suggested that the subject deserved some attention in the popular press. The many issues surrounding women's participation in science—as students, practitioners, and consumers—are widely discussed within women's academic circles, but seem unavailable to young women outside of the academy and even to young women scientists, as well as to the general public. The subject is important enough to receive a more accessible treatment.

As a member of the last couple of years of the baby-boom generation, I've been in a position to observe the social and economic transformations brought about by the masses of women entering public life before me, as well as to view the entry of younger women into their years of independence. This is a source of endless fascination and has led me to question time and again my beliefs and assumptions about women's natures. When the opportunity to look into how women viewed their experience of science presented itself, I rolled all of my preconceptions into a set of questions that were promptly challenged by just about every person I spoke with on the subject. This book was born of some fascinating conversations, with women as well as men. Its creation was not without a certain measure of surprise and a good dose of incredulity, as the notions I held about women and their role in the sciences were deconstructed and debated one by one.

I hope that readers will put aside their own assumptions and listen closely to what is being said by the scientists in these pages. If the book serves a purpose, it is to provide a vehicle to open dialogue and to offer a rational argument for positive action.

Mary Morse

Minneapolis, Minnesota

Acknowledgments

This book owes its existence to a great many people who have given generously of their time and thoughts, foremost among them the scientific women and men who wrote to me in response to my voluminous queries on the Internet. Their frankness and willingness to engage so openly in a controversial subject were essential to the development of the book's premise. Thanks also to the women who allowed me to interview them at length, whose ideas, intelligence, and courage are truly admirable. Special appreciation goes to the women scientists at the University of Minnesota, who allowed me to attend their meetings and question them endlessly.

Thanks to my parents, siblings, and friends for patience, assistance, leads, and encouragement. To Sue Gunderson, whose understanding of balanced lifestyles permitted me to write a book, raise my son, and keep my job, my everlasting admiration. For the loving and flexible child care that surrounded my son Maxwell, thanks are due to Kate, Kim, Annie, Kerry, Eve, and especially Grandma MaryAnn.

My editor Frank Darmstadt provided opportunity and guidance with a delightfully dry wit. Eric Utne, Helen Cordes, Lynette Lamb, and Jay Wall-jasper all deserve credit for nurturing my writing career.

And finally, my love and gratitude go to my favorite scientist and husband Jim, whose own experience in the profession has been remarkably instructive to me.

Contents

Part II Δ Science on Women's Terms

Introduction

Science has become a very large part of our lives. Peoples and nations along the spectrum of development are inextricably linked to scientific endeavor. From what we eat to how we travel, from the products our businesses sell to the health care we receive, science and technology have tremendous impacts on how we define our lives physically, culturally, politically, and economically.

Science has historically been the domain of men. Men have largely determined what gets studied, which technologies are developed, and how science dollars are spent. Though women have always played a part in science, until recently their impact on the discipline's mainstream was minimal. Now that women have begun filling the ranks of science, will its focus, or even science itself, change?

During the fall of 1993, I began asking questions of women involved in the scientific enterprise. At first I worked within a small sphere: talking with women scientists I knew, attending meetings of the University of Minnesota's women in science and technology support group and its women in physics bagel lunches, and contacting women scientists who had recently received press coverage. I asked them how they got started in science and what they thought of their chosen careers. I distributed printed surveys, which were mailed back full of descriptions of lives dedicated to study and research. I phoned women directly and had rollicking, free-form discussions on whatever they had on their minds pertaining to their involvement in science.

From these initial contacts, I began to find that not all was well with women in science. Was a trend emerging or was this a fluke? I began to throw the net wider. I posted questionnaires to several computer discussion groups, including WISENET (Women In Science and Technology Network), FIST (Feminism in/and Science and Technology), AASWOMEN (maintained by the American Astronomical Society for women astrophysicists and astronomers), YSN (the Young Scientists Network), and Systers (for women in computing). The total subscriber pool I reached through these lists was well over 6000, and my surveys were distributed beyond the computer networks by professors and others who took an interest in this line of inquiry. As responses came back, I considered their implications, formulated new questions, and sent fresh sets back out on the Internet. The research I conducted in this manner was far from scientific. I did no sophisticated statistical analyses of responses or respondents. The "results" I came up with were derived directly from the input of more than 87 women and 46 men in the United States and abroad, a diverse pool of research subjects that would have been nearly impossible to access without using the Internet.

Added to the response from computer networks came information I gathered through phone interviews, through meetings, and in casual conversations. All told, I heard the stories of dozens of women scientists. I spoke with African American women, Asian American women, Latinas, lesbians, married and single women, mothers, grandmothers, girls and women from around the world. My respondents ranged from the gainfully employed to the painfully unemployed, from students to their teachers. In most cases, I asked respondents to tell me what it was about them—their race, their class, their age—that *they* felt most defined their experience as a scientist. This book, while including the viewpoints of scientists outside the United States, focuses primarily on the US science system, with which most respondents were familiar.

I approached this subject as a science outsider, which gave me a measure of openness as I began doing the research for this book. I had few preconceived notions about what women scientists would say about their field, other than my expectation to find things vastly improved for them in the last couple of decades, to parallel the gains made by women in other professions. My sense of openness was invaluable, as the communications I began having with women scientists took on a far different tone than I had initially anticipated. The women I spoke with were quite willing or even eager to tell their stories, which emerged with conviction, honesty, and concern. The stories, however, were not pleasant ones. After hearing so many troubling tales, I found myself wondering just how I could write a book about women in science that told the truth about women's difficulties in and with the field without falling into the trap of discouraging young women and girls from embarking on the scientific career path. What resulted is a description of women's science experience today from grade school through career, with a wide variety of viewpoint and opinion coming directly from scientists to the reader.

Chapter 1 offers the premise that women don't always find science relevant to their lives. A brief history of women's scientific education and involvement is provided, followed by a look at the sciences through feminist eyes. The chapter concludes by examining some areas where science has failed to address women's needs.

Scientific training, from kindergarten through graduate school, is addressed in Chapter 2. The external influences that can make or break a young woman's science career are addressed, including peer and family pressures, role models, and the effects of the images of women scientists broadcast through the popular media. For young women who survive the scientific preparatory gauntlet, a tough job hunt awaits. The cruel reality of today's scientific job market is described in depth, including background on just how this dilemma evolved.

Chapter 3 attempts to gauge the "difference" factor in science: Do women think about or do science differently than men? The key themes of the chapter—scientific competition, cooperation, intimidation, and ethics—were all very much on the minds of my respondents. A wild ride is in store for the reader: not all members of each gender are in agreement about these subjects.

The working culture of science is discussed in Chapter 4. What do women like about scientific careers and what do they loathe? Is doing science worth trading in one's personal life? In this chapter, scientists look for a balance between the work they love and the lives they lead.

Chapters 5, 6, and 7 bring women scientists and science policymakers directly to the page during extended interviews. In these chapters, women present their arguments, describe their lives, make fascinating predictions, and address the dilemmas they face as women scientists. The tone of this section is conversational, even "chatty," as one reader described it, a fact intended to retain the veracity of the subjects' experience. By leaving their words intact, I hope to underscore the similarities among women's science experience and lend credence to the nearly unbelievable stories that women tell.

Chapter 8 provides an abbreviated list of actions that may be taken to improve women's experience of science, garnered from my own observations, from the suggestions made by my respondents, and from research results of organizations such as the American Association of University Women. While this chapter attempts to round up the changes that women would like to make to science, many other ideas lie within the main body of the book in the interviews and direct quotes included throughout each chapter.

There is a growing number of books on women and science, on feminism and science, and about the history and philosophy of science. What is not present is a discussion of science by women scientists themselves, in their own words. This is the book I have attempted to provide. I honored my respondents' appeals for anonymity. The fact that it was requested at all reveals much about the risks women take when criticizing their field. Of course, not all respondents asked to remain anonymous, but in order to maintain the focus on their words and not on their risk taking, all but a few women are identified by pseudonym. Where possible, descriptive characteristics are provided that reflect the respondents' ages, backgrounds, races, fields, and professional activities.

I bring to discussions of working culture and careers a distinct viewpoint that the reader will doubtless take note of and that many may find discomfiting. The values that Western society has placed on economic activity, in my view, leave much to be desired. I discuss this at length in Chapter 4, where we see full-time scientific careers consuming literally all of women's and men's lives. While I appreciate the work ethic and wholeheartedly support gainful economic pursuits for both women and men, it is clear to me that US society has completely missed the point of technological advancement and industrialization. Rather than looking about and finding well-rounded workers freed of backbreaking labor and crippling disease, I see a society where people are crushed by consumption, hurrying from home to office and back again, leaving family, community, and leisure in the lurch. My chagrin with the US system was compounded when several scientists described non-US cultures, where fruitful professional commitments appear to be balanced with family and personal activities.

Readers may be amazed to find that I did not set out to unearth the most disgruntled women in science; in fact, even those who told completely unnerving stories about their experiences had very positive attitudes about much of their work—a feature that I hope shines through the gloom in these pages. Women do not want to abandon science; they want it to change. To the extent that women remain in the field, gaining stature and influence, they will, along with the men who share their views, move their agenda for science forward.

I

Women and the Scientific Enterprise

1

From Isaac Newton to Ecofeminism
Bringing Women's Relevance to Science

From the perspective of women's lives, scientific rationality frequently appears irrational.

Sandra Harding,[1]

Science as a Male Pursuit

Several years ago I wrote a short magazine article about women and science. In it I stated that science was historically a male business, having been shaped from Newton's day to the present by its male practitioners. I received a barrage of mail in response to the article, most of which pointed out that women have indeed been practicing science, but have been excluded from established scientific circles throughout history. Because of their exclusion, writers reasoned, women's legitimate contributions to science were either refuted, co-opted by men, or most often simply ignored. The letter writers referred to several recent books on the history of science that explored women's roles in the field, and they cited the work of illustrious female scientists, such as Marie Curie and Jane Goodall. My critics were sure that I hadn't done my homework; that had I looked a bit more closely I would have seen that women had made significant contributions to science. The overriding point seemed to be that women have effectively shaped the course of science from their unacknowledged positions behind-the-scenes.

My detractors were partially right. Of course, women have been scientists all along. There are far more women scientists in history than one would imagine if one relied on the scant handful of well-known figures that appear in most textbooks. Londa Schiebinger, professor of history and women's studies and director of the Penn State Women in the Sciences and Engineering Institute, presents a comprehensive history of women scientists in her book *The Mind Has No Sex? Women in the Origins of Modern Science*. Schiebinger notes contributions from women across the sciences, such as the significant 14% of all German astronomers between 1650 and 1710 who were women.[2] Perhaps part of the "invisibility" problem is that the activities that many women have engaged in throughout history, while scientific in nature, are not formally called "science." Because of their subordinate position in the Western social hierarchy, women's activities were decidedly second rank—at least until men took an interest and revised the nomenclature. Philosopher Ruth Ginzberg,[3] assistant professor at Wesleyan University, has written widely about women's issues, and says:

> In searching through women's activities outside of those that have been formally bestowed the label of "Science," I have come to suspect that gynocentric science often has been called "art," as in the *art* of midwifery, or the *art* of cooking, or the *art* of homemaking. Had these "arts" been androcentric activities, I have no doubt that they would have been called, respectively, obstetrical *science*, food *science*, and family *social science*. Indeed, as men have taken an interest in these subjects they have been renamed sciences—and, more importantly, they have been reconceived in the androcentric model of science.

It was my critics' second point—that women have had significant influence on scientific endeavor—that caused me the greater consternation. While there is no doubt that women have had some effect on science, to insist that our impact on the field in any way approaches that of men is wrong. After several hundred years of dominating science, men have thoroughly defined the culture and the rules of the profession. Science is a specific way of looking at the world and its inhabitants; a systematized method to understand the complexities of nature. Those who call themselves scientists have learned its methodology and agree, generally, to its self-imposed rules. The field is self-regulating. The rule makers are those scientists who receive advanced degrees and go on to establish themselves in research or academia, usually both. Scientists impose an inwardly directed culture by speaking a highly developed and rather arcane language. They invite up-and-coming colleagues to join cadres of professional societies. This insular culture serves to perpetuate itself in a male model. Women have simply been out of the loop in too many numbers for too long to have made a defining mark either on the way science is practiced or on its advances.

To their credit, historians are beginning to acknowledge women's role in shaping the world beyond domestic chores and child rearing. Of course, women have made important scientific contributions, often forced into doing so in non-traditional ways, even so "nontraditional" as having their work appropriated by others. In her chapter, "Constraints on Excellence: Structural and Cultural Barriers to the Recognition and Demonstration of Achievement," in *The Outer Circle,* sociologist Cynthia Fuchs Epstein[4] writes:

> Some women researchers simply had their contributions plundered by men who walked off with the prizes. In England, Rosalind Franklin's x-ray spect[r]ograph evidence of the double helix structure of DNA was pilfered by Crick and Watson, who went on to publish the proof of its structure and become Nobel laureates. Jocelyn Bell Burnell's discovery of pulsars was attributed to her senior co-workers, who carried off the Nobel Prize. Controversy about the importance of their contributions still surrounds the cases of Franklin and Burnell, but it is clear that neither woman found their work environment as hospitable as did their male colleagues.

To suggest that women have played a role in scientific inquiry that in any way approaches that of men's is revisionism in its most naive and damaging form, which serves not to convince of the value of women's activities, but to diminish the possibilities for women's future contributions. Women have been kept out of the system en masse. The system that they are now entering was created by men, was populated by men, and in most cases continues to primarily serve the ends of men. The prospects for women's scientific achievements when they're allowed in the game are incalculable, and the changes science will undergo under women's influence will range from subtle to complete. In this chapter, we will consider some of the reasons why women may see science differently than men and we will examine why some feminists are finding the field so ripe for analysis. We will also look at the concepts of feminine and feminist sciences and consider their merits and possibilities.

Feminism and the Feminist Critique of Science

Feminist scholars are ultimately responsible for illuminating what contributions to science women have made. Their efforts to rewrite history books in order to include the history of women alongside that of men have revealed the hidden extent of women's scientific involvement and have foreshadowed some of the changes women may impose on the field. In her examination of the possibilities of feminist science, English academician Hilary Rose[5] writes:

> Feminism has rendered visible the exclusion of women from science; it has rescued from the oblivion of male history many of the women who have entered science; it has fostered the coming together of feminist scientists to rebut the claims of biological determinism; and it has recognized the theory and practice of the domination of nature as the specific feature of masculinist, bourgeois science.

From acknowledging women's limited but underplayed scientific contributions to analyzing science as a subjective product of its culture, feminists have truly ignited the debate about the value and future of Western science. They do so for the most part from the fringes of the field, as of course not all feminists are interested in science, and certainly not all scientists are feminists. Feminist critics of science are taking on an extremely self-serving, highly retributive, and enormously entrenched culture; to do so as an insider takes a large measure of courage or an equal measure of professional security.

One woman who has succeeded in combining a respected scientific career with active feminist analysis is Evelyn Fox Keller, professor of the history and philosophy of science and director of MIT's women's studies program. Keller, whose background is in theoretical physics and molecular and mathematical biology, has written numerous essays on women, feminism, and science, and is perhaps most well known for her book, *A Feeling for the Organism: The Life and Work of Barbara McClintock*, which described the enduring difficulties faced by a brilliant woman who did not fit the mold of male science culture. Despite being dismissed by some as an eccentric old woman, McClintock went on to win the Nobel prize for her work. Keller[6] shows us how difficult it might be for a woman scientist to publicly embrace the idea of a gendered, feminist science:

> For women who have managed to obtain a foothold within the world of science, the situation is particularly fraught. Because they are "inside," they have everything to lose by a demarcation along the lines of sex that has historically only worked to exclude them. And precisely because they are rarely quite fully inside, more commonly somewhere near the edge, the threat of such exclusion looms particularly ominously. At the same time, as scientists, they have a vested interest in defending a traditional view of science—perhaps, because of the relative insecurity of their status, even more fiercely than their relatively more secure male colleagues.

There is no doubt that there are feminists working within science. Harvard biology professor Ruth Hubbard[7] agrees with Keller in that feminist scientists can become invisible:

> Of course it is difficult for feminists who, as women, are just gaining a toehold in science, to try to make fundamental changes in the ways scientists perceive science and do it. This is why many scientists who are feminists live double-lives and conform to the pretenses of an apolitical, value-free, meritocratic science in our working lives while living our politics elsewhere. Meanwhile, many of us who want to integrate our politics with our work, analyze and critique the standard science, but no longer do it.

In researching this book I encountered a range of reactions from the women I spoke with in response to questions about feminist and feminine science. Most female scientists in their 20s were only vaguely aware of the feminist discussion about science, unless they had taken a women's studies or history of science class, in which case they were quite conversant. Older women were more aware of the discussion and often were able to cite one or two articles or books they had read that addressed the issue. A smaller group was well versed in the literature of feminist criticism and was actively engaged in relating feminist concepts to their own scientific careers.

To scholars and women scientists who have been introduced to the idea, the concept of feminist science—a science whose aims and practice are informed by feminist values—is relatively well understood. To discuss the question of whether there's a "feminine" science, however, one needs first to agree on a definition of femininity. Here we tread on a cultural sponge, soaked with assumptions from all ethnic groups, social classes, ages, and every other cultural division imaginable. While some groups value women's strength or independence, the femininity of the white, dominant culture involves beauty, passivity, nurturance, intuition, and subordination. Because the culture of the white upper classes is the culture of science, for a feminine science to exist today it might very well exhibit some of the same traits.

Women interpret the concepts of "feminine" science, "women's" science, and "female" science in highly personal and subjective ways. Even though some writers are attempting to define the terms, women scientists are all over the map in their own definitions and comfort levels with the ideas. Some women I spoke to were drawn to connect the terms "feminine" and science, and these women often fell into the camps of the "difference" feminists or ecofeminists, both of which are described later in this chapter. In other cases, when I asked women scientists what they thought of "feminine" science, especially if I had provided no lead-in or other context to the question, I was charged with naiveté, a lack of education, or being a traitor—just for asking! These women simply could not relate what they believe are artificially imposed, often undesirable gender behaviors with real differences in scientific endeavors, and the recurring question of whether there is a feminine science was quite vexing to them.

Whether or not women want to accept the concept of femininity that is foisted on them, some scientists are actively encouraging the opening of science to what have been considered feminine qualities. An example is the emerging field of chaos theory. Chaos describes the dynamics of physical systems whose behavior defies prediction by the conventional, deterministic laws of Newton, such as the atmosphere, a population of organisms, or a dripping water faucet. The classical approach to these problems is reductionist to the core: It assumes that if all the minute starting conditions are known, the future behavior will unfold in a linear, predictable fashion. Chaos, on the other hand, offers a much more subtle, holistic view of dynamic systems. While often dashing hopes of predicting systems with certainty, chaos reveals clear patterns behind seemingly random behavior. Chaos theory's focus on the nonlinear qualities of nature have been described by such terms as curvaceousness, fluidity, complexity, and attraction—all characteristics attributed to the Western concept of femininity.

Jeffrey Goldstein, assistant professor in the school of management and business at Adelphi University, explained how chaos is unearthing the feminine aspects of science that have been repressed from mainstream science in a paper he presented at the 1992 Chaos Network Conference. According to Goldstein, some scientists are fearful of the way chaos complicates the order and predictability of their work, so much of which is based on finding and wiping away differences or deviations. Goldstein is enthusiastic about the potential of nonlinearity in science: "The great thing about this return of the repressed is that it is enabling science and mathematics to explore many more aspects of the real world that are better exemplified by irregular and curving boundaries, turbulent movements, seemingly random behavior, and complex and convoluted dynamics."[8]

A large body of literature has been written by feminist intellectuals that examines the historical and cultural bases of Western science. To thoroughly lay out all of the arguments and to discuss the points and counterpoints of various thinkers would take far more space than allowed for in this book. What I'll attempt to do here is to describe a part of the framework of the feminist critique of science, illustrate some basic arguments, and expand on those that are currently attracting the most attention among feminist thinkers. Throughout the book, and particularly in the expanded interviews in Chapters 5 through 7, women will address feminist concepts from their own experience.

As noted earlier, not all women scientists consider themselves feminists, even though it's fair to say that any woman undertaking a scientific career has benefited from the women's movement. Feminism is a sociopolitical movement to some and a state of mind to others. The feminist movement and its outgrowths have come under heavy fire of late from critics. Women's studies programs are being assailed (by women) in such progressive publications as *Mother Jones,*[9] and according to journalist and author Susan Faludi, a backlash against women's emancipation is going on in the media and elsewhere.[10] Popular women writers today include Katie Roiphe, who, in her book, *The Morning After: Sex, Fear, and Feminism on Campus,*[11] scolds her peers for being inappropriately hysterical about sexual assault on campus; Christina Hoff Sommers, professor of philosophy at Clark University, who in her book *Who Stole Feminism: How Women Have Betrayed Women,*[12] accuses newly powerful campus feminists of indoctrinating women's studies students with a pedagogical party line, squelching dissent, and controlling the academy; and baby boomer Camille Paglia, a self-described feminist theoretician whose tirades against "whining" feminists and feminist academics show up so often in popular media that one might assume she is the *de facto* spokesperson for the antifeminist movement.[13]

Angela Ginorio, a Puerto Rican scholar and director of the Northwest Center for Research on Women at the University of Washington, told me a story about a recent course she taught on women and science. So great was her students' fear of being typecast as feminists or of having their scientist colleagues think that they were buying into the feminist critique of science that of the six women enrolled in the course, none had told their advisors that they were taking it and only two had told peers. This certainly suggests that there are antifeminist pressures felt by female science students on college campuses today. How unfortunate that the very movement that made possible their participation as scientists, which is today credited for asking some of the most probing, difficult questions about the values and outcomes of science and which is no doubt going to be the one of the most effective means that women have to effect positive change in their scientific careers, is so feared, reviled, or ignored by so many young women. Let's take a look at the feminist critique of science.

Knowledge as Male

This is perhaps the easiest of the concepts of the feminist critique to grasp, if only because the historical basis for the assertion is applicable, to some degree, to the experience of most scientists living today. With thousands of years of domination over the world's economic and political power structure by men came men's domain over education and the "truth seeking" it comprised. In the Victorian era, women who displayed an interest in education were warned that it could physically harm them, leading perhaps to reproductive difficulties. In 1906, G. Stanley Hall, president of Clark University, sternly warned against young women undertaking rigorous academic studies, suggesting that, "Over-activity of the brain during the critical period of the middle and late teens will interfere with the full development of mammary power and of the functions essential for the full transmission of life generally."[14]

Rather than being allowed to become scientists in their own right, women were grudgingly allowed to act in scientific support roles, such as analyzing data or doing limited fieldwork. According to Harvard's Ruth Hubbard, women's role in science, when allowed at all, was limited to being "wives, sisters, secretaries, technicians, and students of 'great men.'"[15] Rarely were women recognized as the askers of the questions or allowed into any leadership capacities. Those who were full scientific participants were anomalies, often working in their husband's laboratories or in such "safe" environments as women's schools and colleges.

One of these fringe scientists was Elizabeth Knight Britton, who became the preeminent woman botanist during four decades spanning the turn of the century.[16] Married to botanist Nathaniel Lord Britton, Elizabeth Britton continued her prenuptial botanical studies during the course of her marriage, aided significantly by her husband's position as director-in-chief of the New York Botanical Garden. Unlike many other women scientists of her time, Britton had no children or the attendant duties that family placed on women's shoulders. She was free to follow her interests: namely, studying bryophytes, mosses, and related plants. She published prolifically under her own name, attended meetings, took extensive collecting trips, and discussed her work with other leaders in the field. Her status as a woman scientist was certainly not harmed by the fact that both she and her husband hailed from wealthy families. It seems plausible that Britton's success and her acceptance into the scientific fold were made possible as much from her socioeconomic position as her brilliance.

Today's emphasis on teaching women science is the result of a series of events. From the time of their founding, most universities were officially off-limits to women. One notable exception was the University of Bologna, which housed respected professor of physics Laura Bassi (1711–1778). (For a general early history of women in science, see Schiebinger's *The Mind Has No Sex?*[17]) Scientific societies, which began to emerge in the 17th and 18th centuries, pointedly refused to allow women into their ranks, in some cases until as recently as 1979.[18] One historic occasion that opened up scientific inquiry to greater numbers of women was the growth in popular publishing. Author and ecofeminist Carolyn Merchant[19] shows how capitalism freed wealthier women from the mundane tasks of household production, leading to the opportunity for women to begin spending leisure time with books and outside interests:

> The attempt by bourgeois entrepreneurs to capture a wider female audience lay behind the publication of scientific serials for city and country ladies of all social levels . . . The annual *Ladies Diary* begun in 1704 by John Tipper taught science and posed difficult mathematical and astronomical questions for its audience to solve. The semiweekly *Free Thinker*, published from 1718 to 1721 . . . featured articles on natural history for the "philosophical girl" who "did not aspire to masculine virtues" but was above female capriciousness.

Interest in science continued to grow among the upper classes throughout the 17th and 18th centuries, but is was not until the 1880s that women had finally gained admission to colleges and universities in Europe and the United States.[20] By 1920, women were entering PhD programs, thereby acquiring the professional credentials to do science along with men. However, the proportion of women science faculty and PhDs fell from 1930 to 1960.[21] The causes for the decline appear to include a determined backlash from male scientists to re-exclude women[22] and being crowded out by the huge post-World War II influx of male students under the GI Bill. It was not until the 1970s that the number of women scientists in academe again reached the levels of 50 years earlier, aided no doubt by the passage of the 1964 Civil Rights Act, which barred discrimination on the basis of sex in education and employment.

The Contextual Nature of Scientific Inquiry

An important argument of some feminists is that science, like any other human endeavor, is grounded in the dominant cultures within which it is practiced. This argument is generally accepted in its first precept: That the questions asked by scientists are formulated inside their own experience. When scientific fields were almost exclusively populated by men (here I must point out that several fields still are: among them entomology, physics, and engineering), one might expect that scientific activity would be useful to, interesting to, or applicable to the male experience of the world. I would not suggest that any one "male" experience is common to all men; surely the vast differences in class, education, race, ethnicity, and power that exist among the world's men are as diverse as those of women. But given the fact of men's numerical dominance in science, it follows that the cultural traits men bring to the laboratory and classroom will inform their research interests and their interpretation of results. This makes the practice of science inherently political and detracts from the concept that science is a pure pursuit. The political science versus pure science debate is related to feminist empiricism, a concept that will be considered later in the chapter.

Carol Cohn, professor of women's studies and sociology at Bowdoin College in Maine, spent time with men she called "defense intellectuals": professionals in the nuclear weapons field. Cohn related an example of male cultural dominance in science in her essay "'Clean Bombs' and Clean Language."[23] Under the heading, Men in Ties Discussing Missile Size, Cohn relates that she was under the naive impression that the feminist critique of science would have affected the men she encountered. Instead, she found a culture not only dripping in clever euphemisms to tone down the potentially horrendous and catastrophic outcomes of their work, but rife with male references to a sexual subtext. Cohn describes some of the lectures she heard:

> . . . lectures were filled with discussion of vertical erector launchers, thrust-to-weight ratios, soft lay-downs, deep penetration, and the comparative advantages of protracted versus spasm attacks—or what one military adviser to the National Security Council has called "releasing 70 to 80 percent of our megatonnage in one orgasmic whump." There was serious concern about the need to harden our missiles, and the need to "face it, the Russians are a little harder than we are." Disbelieving glances would occasionally pass between me and my one ally—another woman—but no one else seemed to notice.

The terms of arms production that Cohn encountered were "racy, sexy, snappy," and once learned, became part of an enjoyable and exciting discourse that she shared with the men around her. In this way, she illustrates how a dominant group can define the culture of a scientific or technical field and how the practitioners that follow, even though they may not be native to the dominant culture, can adopt, defend, and even enjoy its traits. This may relate to the unbending loyalty some women (and men) express to the science status quo: Even though they may hold reservations about the direction of research or the applications of their work, they have spent so much time learning the scientific "language," that once fluent they become its avid defenders. Cohn concludes her essay by pointing out the difficulties faced by feminists and others who seek "a more just and peaceful world," stating that the dominant masculine voice speaks so loudly in Western culture that it drowns out all others.

In her article "Can There Be a Feminist Science?" Rice University philosopher Helen Longino[24] distinguishes two types of values that she feels are relevant to science: constitutive values, which are the internal values that govern the scientific method, and contextual values, which comprise the cultural, personal, and social contexts in which science is practiced. Longino refutes the much-proclaimed principle that the two value forms are completely distinct in science; arguing rather that a distinction cannot be maintained. She contends that one cannot completely eliminate the role of value-laden assumptions in the scientific method. Longino elaborates:

> This is not to say that all scientific reasoning involves value-related assumptions. Sometimes auxiliary assumptions will be supported by mundane in-

ductive reasoning. But sometimes they will not be. In any given case, they may be metaphysical in character; they may be untestable with present investigative techniques; they may be rooted in contextual, value-related considerations. If, however, there is no a priori way to eliminate such assumptions from evidential reasoning generally, and, hence, no way to rule out value-laden assumptions, then there is no formal basis for arguing that an inference mediated by contextual values is thereby bad science.

According to some feminists, science's self-anointed "objectivity" is little more than a relativist positioning of the majority culture of scientists. Longino's assertions are supported by Sandra Harding,[25] a professor of philosophy and director of women's studies at the University of Delaware who has written widely about feminism and science. Harding develops the concept that an objective, value-free science is improbable, given that researchers tend to share social values and interests:

> If the community of "qualified" researchers and critics systematically excludes, for example, all African Americans and women of all races, and if the larger culture is stratified by race and gender and lacks powerful critiques of this stratification, is it not plausible to imagine that racist and sexist interests and values would be identified within a community of scientists composed entirely of people who benefit—intentionally or not—from institutional racism and sexism?

Sciences practiced within a feminist context exist now, particularly in the biological and social sciences. An example is the recent work of evolutionary biologist Margie Profet on menstruation. Profet's ground-breaking look at an essential biological process turns on its head the accepted tenet that menstruation occurs only to prepare the uterus for fertilization. Profet's work, for which she was awarded a $225,000 MacArthur Foundation grant, is of the type that may never have come out of nonfeminist science. Her hypothesis, that the menstrual flow aids the body in cleansing the potentially harmful products introduced into women's reproductive tracts through sperm, is a totally new way of considering a biological process.[26]

Even though work such as Profet's shows the promise of expanding the range of scientific participants (particularly since she holds no academic credentials in her field—just two bachelors degrees, one in physics and another in political philosophy), it is open to question whether all fields would be equally affected by increased involvement of women. In the physical sciences, where little if any research involves human information, it would be difficult for the feminist critique to affect anything beyond relationships within the lab. Another point made by Harding is that research in fields such as physics and chemistry is carried out on an international level as opposed to social research projects that are typically locally organized and more susceptible to feminist influences.[27] In the end, it's far from impossible that the scientists who are pushing the boundaries of what is acceptable scientific thought, such as feminist thinkers and chaos theorists, may find their work bringing them together. Their combined focus on culturally unbound, nonlinear research may help these ideas gain mainstream credibility.

The "Difference" Factor

Some feminists believe that women's science experience differs from men's due primarily to gender variances. Some of the attributes that women are said to share are described as caring for and relating well to others, thinking intuitively, exhibiting an openness to emotionality, and tending to cooperate rather than to compete. The basic premise of "difference" feminism, which is also known as "cultural" or "romantic" feminism, is that such "women's" attributes, though long relegated to the unimportant domain of the weaker sex, are in fact at least an equal and perhaps a much more highly developed and valuable set of characteristics than traditional male traits such as dominance and stoicism. A notable proponent of difference feminism is Harvard professor of human development Carol Gilligan, who in her 1982 book, *In a Different Voice,* described several studies in which subjects were interviewed about their "conceptions of self and morality," and about their "experiences of conflict and choice." Gilligan laid out her rationale for what she believes to be the value in women's gender-related traits. She makes it clear that she is not making a case for biological determinism, and that the different traits held by women are not only not absolute, but "arise in a social context where factors of social status and power combine with reproductive biology to shape the experiences of males and females and the relations between the sexes."[28] If one accepts Gilligan's brand of difference, then one can easily see that not only women will exhibit differences, but that differences can be found emanating from all along the racial, cultural, and social spectra.

Another group of difference enthusiasts espouse the idea that humans exhibit genetically predetermined traits or temperaments based on inherent sexual, racial, and even ethnic differences. It follows that some of these differences would result in female behaviors and characteristics that are inborn and distinctly different from those innate to men. This theory, sometimes known as biological determinism, has a dubious history and reputation among scientific cognoscente, yet the field's adherents are not without luminaries. In spite of scholarly skepticism of biological determinism, researchers continue to search for and publish new evidence that they (sometimes reluctantly) interpret as biological impetus for the way men and women think and behave.

An example is Doreen Kimura, professor of psychology at Canada's University of Western Ontario and winner of the 1992 John Dewan Award for outstanding research from the Ontario Mental Health Foundation. Kimura has written about a much-studied area—the role of sex hormones in determining cognitive differences between men and women.[29] Her research has shown evidence that early exposure to estrogens and androgens (as early as in the prenatal or neonatal stage) may affect brain development in children, essentially fixing different patterns of cognitive ability along sex lines. Kimura's evidence includes results of problem-solving tests, where women outperform men on perceptual speed, verbal fluency, fine motor coordination, and mathematical calculation, but are less capable in spatial tasks, target-directed motor skills, and mathematical reasoning. Kimura believes that the effects of early exposure to sex hormones are distributed along a range in males and females, with some individuals lying toward the endpoints of the distribution. However, her work indicates that having high levels of particular hormones after the critical early-life exposure does not necessarily lead to particular proficiency in tests, and she points to other research that shows routine hormonal fluctuations affecting cognitive patterns throughout men's and women's lives.

Just how does this neonatal dose of hormones play out in adulthood? Kimura believes that individuals might exhibit specific occupational interests and abilities as a result of hormone effects. In her article, "Sex Differences in the Brain," Kimura concludes that,

> I would not expect, for example, that women and men would necessarily be equally represented in activities or professions that emphasize spatial or math skills, such as engineering or physics. But I might expect more women in medical diagnostic fields where perceptual skills are important. So that even though any one individual might have the capacity to be in a "nontypical" field, the sex proportions as a whole might vary."

Kimura is not alone in her emphasis on biology over environment. In her article "How We Become What We Are," Winifred Gallagher[30] examines the work of scientists such as Harvard developmental psychologist Jerome Kagan, who, she reports, used to emphasize the role of environment on human behavior. Kagan avows that his "training, politics, and values" previously led him to "mute the power of biology" in favor of environmental factors. Now, Kagan reluctantly admits, "I have been dragged, kicking and screaming, by my data to acknowledge that temperament is more powerful than I thought and wish to believe."

Biological determinism, as stated before, has a long and unfortunate history in science, where it has been used to support racial superiority theories, the rightfulness of the sexual division of labor, and many other bizarre concepts.[31] There seems to be no way that even feminist supporters of biological determinism can predict, much less control, the outcomes of this debate. Once allowed into public discourse, the biological bases of sex differences could be used to justify a slate of ultraconservative policies, putting women squarely back into the kitchen—barefoot.

It's important to point out that like biological determinism, difference feminism is a much-debated topic and is reviled in certain feminist quarters. Feminist writer Katha Pollitt[32], associate editor of *The Nation*, observes that,

> Although it is couched in praise, difference feminism is demeaning to women. It asks that women be admitted into public life and public discourse not because they have a right to be there but because they will improve them . . . why should the task of moral and social transformation be laid on women's doorstep and not on everyone's—or, for that matter, on men's, by the you-broke-it-you-fix-it principle . . . By promising to assume that responsibility, difference feminists lay the groundwork for excluding women again, as soon as it becomes clear that the promise cannot be kept.

I attempted to gauge the level of support for difference feminism among the female and male scientists with whom I spoke. Might women, because of some inherent feminine trait, "do" science differently from men? The vast majority of the scientists I spoke with or who responded via the Internet felt that the basic motivations and methodologies used in science are the same for both men and women. Since I guaranteed anonymity to anyone who requested it, I am doubtful that this response was provided out of fear. A more obvious reaction that I heard was disgust. Women scientists believed that once again, even though it was couched in the noblest of intentions, women were being singled out by difference feminists as unequal, and that by being placed apart from men they would necessarily be judged differently and likely inferior in ability to men.

Women who have made it in science want least of all to be patronized because of even a hint of gender-driven behavioral or intellectual difference.

Acknowledging differences is clearly dangerous for contemporary women scientists. If research such as Doreen Kimura's continues to support biological bases for the way we think and reason, then wholesale career tracking might become a reality along sex lines, even though it's quite obvious that individuals of both sexes show significant variations in ability and temperament. Yet should the research hold up and we become aware of women's particular strengths, it would be an opportunity to encourage girls to enter the areas of science and other careers in which they might naturally excel, rather than expect them to excel in fields where they may have some measure of natural disadvantage. Women and men who stand in total opposition to the consideration of research on differences may be doing us all a disservice.

Feminist Empiricism, Feminist Standpoint Theory, and Feminist Postmodernism

While they may not have known about the feminist epistemologies that described their opinions, many of the scientists with whom I conversed explained their view of science in terms that fell fairly neatly within one of three constructs proposed in the mid 1980s by Sandra Harding and now widely acknowledged and discussed among feminist academics. The first of these, *feminist empiricism*, argues that the problems feminists find with science do not pertain to the internal guidelines of the sciences themselves, but rather to differences in competence found among scientific practitioners. This is a "good science–bad scientist" argument; it states that a scientist whose work is tainted by racist, sexist, or otherwise twisted or incorrect assumptions is simply practicing "bad" science. Under the feminist empiricism model, science's rules and methodologies lead to true and objective results when practiced correctly; when research is carried out in the purest scientific fashion it should lead to universal truths, applicable and acceptable to everyone. Harding finds that feminist empiricism is less threatening to the practice of science than other feminist theories. She says, "Feminist empiricism appears to leave intact much of scientists' and philosophers' conventional understanding of the principles of adequate scientific research. It appears to challenge mainly the incomplete practice of the scientific method, not the norms of science themselves."[33] The tenets of feminist empiricism were supported by many of the women scientists

I spoke with, and the epistemology was notably strong among the men who responded to my questions. It clearly is seen as a noble criticism of science, and it places guilt for unfortunate or ridiculous outcomes on unworthy practitioners, rather than on the larger science culture or science itself.

Feminist standpoint theory differs from feminist empiricism in that it acknowledges the external position from which women have experienced the sciences, and it embraces this perspective as women approach the field with questions and research priorities that reflect their own experience. Rather than accepting the basic doctrines of science, feminist standpoint theory seeks to examine the foundations that these doctrines stand on by examining the universe from the feminist standpoint. For example, one of the goals of standpoint theorists is to describe the social and political hierarchies of modern science, which might include looking at gender and racial makeup of scientists, class issues, issues of ethnicity or nationality, sexual preference, or others as they relate to science culture. From these descriptions follow questions about the relationship of science, historically dominated by white males, to those who have until recently simply experienced its effects, namely women, nonwhites, non-Western peoples, and nonscientists. It is feminist standpoint theorists who make scientists the most nervous; their views are considered radical among nonfeminist circles. It is not obvious that science would survive in any form resembling its current one if it were to be restructured along the lines advocated by feminist standpoint theorists.

The third epistemology coming out of feminist inquiry is known as *feminist postmodernism*. This theory advances the idea that there may in fact not be a universal experiential reality; that the dozens of identities that humans claim, such as race, sexual orientation, IQ, gender, class, political leanings, and others all confuse and make impossible science's quest for universal truths. Rather than throw out science, postmodernists would deconstruct all existing scientific "knowledge" based on the historical factors and contingencies in place at the time the knowledge was gained. We would end up with perhaps several interpretations of the same events or objects. Future researchers would carefully document their own historical context, acknowledging their own relative positions in the culture, rather than claiming objectivity.

All three of these feminist theories are at the center of considerable debate within the academy. Adherents to feminist empiricism, on the one hand, may believe so wholeheartedly that the only problem with science is its openness to being badly practiced that they disallow any deep questioning of the ethics, goals, and functions of science. On the other hand, feminist standpoint theorists have been accused of ignoring the needs of practicing women scientists, focusing on what the critics claim is just another relativist position (the feminist position) rather than on practical matters such as education and hiring reforms. Harding[34] is concerned about the disagreements among feminists about the nature of the science problem, and notes:

> It is distressing that an apparent consequence of the success of feminist criticisms of this field would be to alienate women from entering it. . . . It is not a purpose of the feminist criticisms of literature that women should stop writing, or of feminist criticisms of the social sciences that women should abandon attempts to understand the social world. How bizarre if it is an outcome of the purportedly most radical feminist science criticisms that women should give up trying to understand the natural world.

To add another layer of complexity to this discussion, one can acknowledge the positions of the critics of the feminist critique of science. Noretta Koertge,[35] professor of the history and philosophy of science at Indiana University, published an editorial in *The Chronicle of Higher Education* that questioned those promoting what she terms the "new feminist agenda." A defender of the scientific method and its built-in objective checks and balances, Koertge wonders what was wrong with the "traditional" feminist analysis that simply required barriers to women's participation in science to drop. She fears that feminist critics have so impugned the motives and value of science that, "the feminist credentials of any woman who does manage to succeed in science today are open to question."

Feminist postmodernists are lambasted by Steven Goldberg,[36] chairman of the sociology department at City College in New York. In an article in the conservative magazine, *National Review,* Goldberg reacts to the idea that universal scientific truths are impossible to define:

> Science leaves far less room for differing views of truth: someone who believes that gravity is such that when he lets go of a bowling ball it will float gently upward is simply incorrect, and someone who believes it will fall to earth is correct. This is validated by correct prediction and by the painful, swollen foot that accompanies the incorrect prediction.

It remains to be seen whether any of the feminist critics of science will see their particular stance gain widespread support. Certainly one of the hallmarks of women's impact on the profession is that feminist criticisms are being discussed by mainstream scientists at all. The critique has touched a nerve, which at the very least may mean that scientists will be even more challenged to follow rigorously objective methodology and at most will force a reassessment of the entire concept of objectivity. It is far too early in the philosophy's life to judge outcomes; but with the basic outline of the feminist critique of science before us, it will be easier to observe the feminist influence on the scientists whose ideas are introduced later in the book.

Is Science Relevant to Women?

Science has come under a good deal of criticism of late, and not just for the exclusion of women. Some observers believe that it has actively sought to control, subjugate, or otherwise hold power over nature. In this section we will take a look at some of the outcomes of science that have led to such criticism.

Ecofeminists are among the most strident of science's critics. Carolyn Merchant, professor of environmental history, philosophy, and ethics at the University of California, Berkeley, thoroughly probed the evolution of modern science and has revealed a historical continuum during which she believes science turned its back on the very earth it claims to be interested in. In the process, Merchant[37] fears that science also closed itself to women:

> In investigating the roots of our current environmental dilemma and its connections to science, technology, and the economy, we must reexamine the formation of a world view and a science that, by reconceptualizing reality as a machine rather than a living organism, sanctioned the domination of both nature and women. The contributions of such founding "fathers" of modern science as Francis Bacon, William Harvey, Rene Descartes, Thomas Hobbes, and Isaac Newton must be reevaluated. The fate of other options, alternative philosophies, and social groups shaped by the organic world view and resistant to the growing exploitative mentality needs reappraisal.

Merchant describes the social and commercial settings that nurtured the development of science, describing how, during the 1600s, the machine became ascendant over the earth. From a culture of taking from nature what was necessary to survive, Europeans began fabricating sophisticated structures and tools and trading in natural resources. She notes that even while early Mediterranean and Greek cultures practiced mining and forestry, they did so at a limited pace, and their societal norms prevented them from overstepping the lines of sustainability. It wasn't until the economy developed along modern lines that the full-bore attack on the earth's resources began. Merchant[38] claims that as the religious doctrine of rightful dominion over the earth was injected into more nature-centered cultures, humans unleashed a powerful onslaught against their home planet and its resources.

> The image of the earth as a living organism and nurturing mother had served as a cultural constraint restricting the actions of human beings. One does not readily slay a mother, dig into her entrails for gold or mutilate her body . . . As long as the earth was considered to be alive and sensitive, it could be considered a breach of human ethical behavior to carry out destructive acts against it.

Merchant is less critical of the "internal content" of science—its accepted conducts and practice—than she is of the cultural norms that have shaped modern science. As mentioned before in the discussion of feminist empiricism, this theme came up repeatedly throughout the interviews I conducted with women scientists and has been widely written about. Women who do science are in love with the profession. They told me of the beauty in its clarity, of the passion in its inquiry, and of the excitement in its newness. If women criticize science, it is more likely to be along the lines of Merchant's arguments; women scientists are troubled that the field they love results in applications that so often treat nature so callously.

Consider the assertion that science has neglected the female. A good way to investigate how science operates with little regard for the feminine is to review some of the products of science. Some may say that this approach is flawed; that science is not in the business of producing "products," but is limited to asking questions and seeking answers. In either case, however, the product or answer affects our experience of the world. There are numerous examples of what we might call sex-biased research, where women's specific biological and health needs are inadequately addressed. These examples will form the basis of my discussion that not all science has served the needs, nor met the expectations, of women.

Women's Health as a Scientific Indicator

When attempting to establish relevance, we might assume that the most immediate interest of any person would be their own physical state. What could be more primary than one's health? From this standpoint, then, we can look at scientific activities that affect women's health and compare them to the activities undertaken on behalf of men's health.

A 1994 committee report from the Institute of Medicine, a private, nonprofit health policy organization functioning under congressional charter to the National Academy of Sciences, concluded that women could not benefit fully from clinical research until they were included as research subjects.[39] The committee noted that women have been summarily excluded from health research for years. For the most part, this exclusion stemmed from researchers' fears about adverse outcomes to participants' child-bearing. In fact, no pregnant woman is currently allowed to participate in a clinical study that is in any way supported by the federal government. It is true that any participant in invasive health research is exposed to risk, and that women of childbearing age must be completely informed of potentially hazardous outcomes. But, as health activists and others point out, by excluding any group from research, one deprives them of any potential benefits to be learned from their participation.

Overprotectiveness is, on its face, a kind but misguided motive for excluding women from basic health research. Another motive may be at play in some studies. Some researchers do not like working with women because they believe their natural hormonal fluctuations would skew data. Here again, the obvious outcome is that study results may not be applicable to women. In order to address this inequity, a federal law was passed in 1993 that would increase the number of women and minorities in any clinical study sponsored or regulated by the National Institutes of Health, the major federal funder of basic medical research.

Biologist Sue Rosser, in an article titled, "Re-Visioning Clinical Research: Gender and the Ethics of Experimental Design,"[40] explains that priorities in medical research are largely determined by federal funders. Rosser underscores the supposition that since most of the decision makers in these funding organizations are white men, their gender status has "definite effects on the choice and definition of problems for research." She includes as examples the notion that certain diseases, which affect women as well as men, are defined as "male" diseases (such as heart disease—a leading cause of death in older women); and that "research on conditions specific to females receives low priority, funding, and prestige."

It becomes evident that until only very recently have women's health issues been taken seriously by major scientific institutions. This was prompted in part by an unprecedented number of women entering the biological sciences during the 1970s and 1980s. Only when women were able to increase their presence among the ranks of working health researchers, physicians, and health policy decision makers did they begin to make an impact on changing medical priorities, and most would agree that recent progress is only the beginning.

A simple case where women's biology has been inadequately studied is the problem of menstrual cramps, or dysmenorrhea. With the advent of each month's menses, a percentage of women experience pain, ranging from mild and short-lived abdominal discomfort to severe distress that can last for days. Menstrual pain is nothing new, but it would appear that research into the problem is. According to Sharon Golub, in her book *Periods: From Menarche to Menopause,* researchers place the percentage of women who suffer dysmenorrhea at some foggy point between 3 and 70%.[41] Dysmenorrhea is sometimes linked to treatable disorders, such as endometriosis, but for many women the cause is unknown. For years, physicians have been unable to prescribe anything more effective than over-the-counter pain relievers and hot water bottles for treatment. Some researchers have even proposed that cramps and other menstrual problems are related to women rejecting traditional role expectations, as Golub paraphrases the research, "Women who do not like being women get dysmenorrhea."[42]

Imagine if a significant percentage of men suffered from regular, excruciating pain involving their reproductive organs from around age 12 until their late 40s. Would they be asked by their physicians to accept the pain and get on with life? Probably not. Would they be told that the reason for their pain was because they rejected masculinity? Decidedly not. More likely there would be a concerted effort to find the cause and possible alleviators of their suffering by health researchers and funders, historically an overwhelmingly male group. It's not difficult to understand why men have not made much effort in this area—they have no direct experience of the problem. Adding insult to injury is the fact that menstruation is considered evil or unclean by many societies—menstruating women being variously forbidden to cook, bathe, visit friends, wash their hair, or even leave a special menstrual hut during their menses. A concerted effort to address menstrual difficulties by the male scientific establishment in these societies seems unlikely.

To look at the health-bias issue another way, consider the efforts now being undertaken to control fertility and reproduction. Some of the most significant strides in female contraception have been on methods such as Depo-Provera, an injectable drug that prevents conception for up to 3 months at a time. While on the surface the answer to many women's dreams, this long-lasting drug therapy is known to cause debilitating and bizarre side effects, such as unexpected uterine bleeding, excessive weight gain, abdominal pain, fatigue, hair loss, and weakness. A suspected link between Depo-Provera and the development of breast cancer led to a 20-year battle between women's groups and the drug's manufacturer, Upjohn, before the drug was approved in the United States in 1992. More recent studies suggest that Depo-Provera may contribute to osteoporosis, a disease that depletes bone mass in older women, leaving bones weak, brittle, and highly breakable.

Observers are concerned that Depo-Provera is being marketed to poor women in the United States and around the world, not so much in an effort to support women's own family planning efforts, but to extend control over their fertility rates through potentially coercive means. In the United States, the National Black Women's Health Project opposed the FDA approval of Depo-Provera, stating that it would create unnecessary risk to African American women, who already suffer from poor health status.[43] What we're left with is a potentially dangerous yet highly effective means to control child-bearing, which is being promoted by public health agencies in poor American cities and by world population groups in developing nations.

It's one thing to offer women a range of safe and effective means to control their own fertility and to provide education and information about reproduction and family planning. It's quite another to promote the fastest route to suppress women's fertility, regardless of its safety level. Disadvantaged women fall victim to the desires of outside interests, including groups such as family planning agencies, which may honestly wish to advance the well-being of women but haven't the money or other resources to get to the issues that directly lead to explosive population growth. It's far easier to distribute injections than it is to redefine the cultural or economic value of large families, or to change men's attitudes about wearing condoms, or to develop an equitable global economy. In the Depo-Provera case, we see science claiming to help women. Whether this is the kind of "help" women really need is highly doubtful.

The flip side of contraceptive research is in the fertility promotion area, which is currently enjoying a boom of sorts. On the face of it, infertility studies and technologies are truly heroic. As anyone who has had difficulty in conceiving or carrying a child might attest: the urge to reproduce, once the notion is in place in one's mind, is incredibly strong. Women willingly submit to complicated, painful drug therapies and surgeries just to have a chance to have a baby.

Is the growing field of infertility research really about helping women?

Any time the scientific community approaches issues of reproduction, their work is rife with societal meaning and moral relativism. The German Nazi Party was famous for its interest in reproductive science. Its support of eugenics, the movement dedicated to improving human "stock" through carefully controlled breeding, is well documented. It's a concept that looks unassailable on the surface: Why would anyone want to ignore practices that could eliminate human suffering from hereditary diseases or defects, could improve the intellectual capacity of humanity, and could prevent the pain and burden of parents who discover their child has a significant flaw?

It is the assumption that one group can identify and even legislate what constitutes a human "flaw" that constitutes the problem with the eugenics movement, a point illustrated by genocidal campaigns throughout history. The technologies that make conception possible are also useful in tampering with normal reproductive processes, bringing in all kinds of ethical and interpersonal dilemmas. In some cases, conception becomes a transaction, where infertile couples and singles purchase the reproductive services of a surrogate who gestates, bears, and then relinquishes her child to the purchasers. These women are often aided in achieving conception with medical techniques, which boost the likelihood of pregnancy, resulting in a less expensive deal for the buyers. Infertility research and product development activities, far from being lauded as useful to women, are being discussed as ways to control women's bodies for scientific and corporate purposes. Women's health activist Gena Corea[44] explains:

> The new reproductive technologies are *not* all about helping the infertile. That is the sugar coating on the pill. The technologies are about controlling women, controlling child production, controlling human evolution. They are also, of course, about making money, setting up corporations which sell women's reproductive services and women's body parts—eggs and wombs. Pharmacrats understand all this. We had better as well. The "morally superficial" among us, as well as those of us who are "low quality life," need urgently to act now in defense of the lives of the women who will be born— or decanted—after us.

When it is no longer difficult to picture women as breeding animals, selling their wombs in surrogacy, science has a problem. When newspaper features on the latest triplets, quadruplets, and quintuplets born of mothers using fertility drugs become commonplace, science has a problem. And when parents check into the sex of their fetus to determine whether to abort the pregnancy before the child is born, science has a problem. When scientists claim that the outcomes of their work, taken out of the laboratory, are no longer their responsibility, then the problem becomes ours. The public's role is to move beyond complacency and to undertake not only to understand what motivates scientists, but to develop standards of ethics with which to govern our own actions and economic activities. Women, recipients of some of science's rawest deals, may be well positioned and well advised to advance initiatives that would increase science's relevance to their lives and well-being.

In the coming chapters we'll look into women scientists' attitudes about the outcomes of their work and how they perceive their role in the lab given their private concerns as parents and breadwinners. We'll talk with women whose moral beliefs are challenged every day by their scientific work and with women who are more than glad to subscribe to the scientific status quo. With more than 20 years of mainstream feminism under our belts, we'll see how women in the worlds of science, engineering, and medicine view themselves and their relationship to their fields and what they foresee in the coming years.

2

Women's Scientific Training and Its Outcomes
All Dressed Up in Lab Coats and Nowhere To Go?

In order to really understand women's experience of science, it's important to consider the factors that shape and inform contemporary women's participation in the field. This chapter begins by directly asking women their motivations for entering the sciences. It continues with a look at the educational and cultural factors that affect young women considering science careers, and concludes with an examination of the current difficulties in the scientific job market. Along the way we'll hear the voices of survey respondents and others who provide meaningful insights into the nature of women's experiences.

On Becoming a Scientist

"How and why do women become scientists?" This question was asked of subscribers to a number of computer bulletin boards used primarily by women scientists, primarily FIST (Feminism in/and Science and Technology) and WISENET (Women in Science and Engineering Network). The answers give us a glimpse into women's motivations, which include an abiding fascination with the natural world and a sense of enjoyment in intellectual challenge. Several women could envision no other possible career path.

According to Carol: "This question is like asking a bird why it chose to fly. I was born a scientist. My interests and my talents all pointed me consistently in that general direction. The decision I have had to make is what kind of scientist I am. Even now, I don't think I've finalized that one. I chose engineering over a natural science for financial reasons, i.e., being able to support myself through school. I chose mechanical engineering because it was the most general and the closest to the physics that I most enjoyed. I later moved to electrical engineering in order to work with a particular faculty advisor. My specialty is neural networks—a blend of mathematics, computer science, computer engineering, information and systems engineering, neuroscience, and psychology. But I continue to closely follow developments in other fields, such as astrophysics and meteorology."

The first person to graduate from college with a science major from either side of her family, Gina is a research assistant in astrophysics. She says, "I became a scientist because astronomy is the only thing that has jumped up and gripped me with both hands. In all my studies, I have never had anything else reach inside my body and take hold of my soul like astronomy. It seems to me that every other field studies the earth and its inhabitants, while astronomy studies everything else. I found that extremely compelling, and the sheer vastness of the things that we do not know is exciting in itself. Basically, we know virtually nothing about the universe, and are making an effort to reduce the amount of knowledge we don't have yet. I became an astronomer because I want to be a part of that effort."

From Marie: "I had the ability, the energy, and the interest. I became [a scientist] because I *could* become one. Of course, the negative aspect of that is that I didn't really like being very many other things. I mean, being a rock musician was great but it didn't exactly pay the bills."

"I am interested in environmental ecology and conservation biology because I feel that time has run out," said Lisa, who believes that environmental protection will occur through study and understanding of natural systems.

Rachel is not reticent in her views. "'Power and influence' is what I always answer to that question," she said. "I'd like to have an academic career and eventually become an administrator to change things from the inside."

"I always wondered how things worked and liked taking my toys apart and putting them back together more than playing with them," said Gwen, a second year grad student in astronomy. "I also spent most of my time building houses and such for my Barbie dolls, rather than playing with them in the normal way. I always loved astronomy and would just stare at the moon as I rode in our family's car, wondering why it would move with us. I would also lie awake in bed wondering about space and what space is held in and get a headache."

A sentence that says it all came from Sarah, a 48-year-old mother of three who recently embarked on her first tenure-track appointment in biological and agricultural engineering: "[Why did I become a scientist?] It was what I was meant to be."

These respondents, and many others like them, convinced me that women become scientists for the same reasons that men do: They are gripped at an early age with wonder and fascination for the natural world; they discover their own facility with math or early science classes; and, perhaps most importantly, no one (or no thing) has stepped in to dissuade them from their interest. Notably few were responses from women who entered science for utilitarian purposes or for overtly political reasons. For instance, no one told me that their sole reason for studying microbiology was to cure AIDS, though several respondents did provide a more general response that science offered them the opportunity to contribute value to society.

The Early Educational Experience for Women Scientists

In this country, a young woman with an interest in a scientific career must wend her way through an educational system that is not well known for sustaining girls' interest in the sciences. The problem of losing potential female scientists during the elementary and secondary years is well documented. One study, conducted by education researchers Samuel Peng and Jay Jaffe, using 1972 data, showed that this loss was directly related to the small number of women continuing on into college science and engineering studies.[1] Nothing has changed in 20 years: In a 1992 report prepared for the National Science Foundation,[2] author Patricia White writes:

> One major factor contributing to women's underrepresentation in the science and engineering work force is that, at any educational level, women do not participate in science and mathematics training to the same extent as do

men. Differences in participation—and interest—in mathematics and science appear first at the elementary and middle school levels. For example, the results of mathematics skills assessments (made at ages 9, 13, and 17) indicate that females' performance starts to lag behind that of males among 13-year-olds (middle school). On science assessments (also made at 9, 13, and 17) females score lower than males as early as age 9.

Although females take almost the same number of years of mathematics and science coursework, they are less likely to take advanced coursework in these subjects. These data, taken together with differences on mathematics and science skill assessments, indicate not only that potential leakages in the science and engineering pipeline are greater for females than for males, but that leakages for females occur very early in their precollege experience.

In 1990, the American Association of University Women, in cooperation with the Wellesley Center for Research on Women, came out with a groundbreaking report on the education of girls, which "challenges traditional assumptions about the egalitarian nature of American schools."[3] *How Schools Shortchange Girls* examined an enormous body of research on the educational experience for girls from preschool through high school, and found that even though girls were entering school exhibiting roughly equal ability to boys, they left with significant deficits in many academic areas. Among the report's findings:

Δ   Girls receive significantly less teacher attention than boys.

Δ   Reports of sexual harassment of girls by boys are increasing.

Δ   The gender gap in math is small and declining, but girls are still not pursuing math-related careers in proportion to boys.

Δ   Girls who went on to take some college math closed the wage gap with men in their fields.

Δ   The gender gap in science hasn't declined and may be increasing.

Δ   Curricula commonly ignore or stereotype females.

Δ   Many standardized tests contain elements of sex bias.

At the end of the report are 40 clear recommendations toward achievement of educational equity for girls.[4] Several relate specifically to science education:

Δ   Educational institutions, professional organizations, and the business community must work together to dispel myths about math and science as "inappropriate" fields for women.

Δ Local schools and communities must encourage and support girls studying science and mathematics by showcasing women role models in scientific and technological fields, disseminating career information, and offering "hands-on" experiences and work groups in science and math classes.

Δ Local schools should seek strong links with youth-serving organizations that have developed successful out-of-school programs for girls in mathematics and science and with those girls' schools that have developed effective programs in these areas.

Educational equity for girls is a hot topic. An anonymous respondent, after reading my questionnaire, suggested that I address the issues of "not providing equitable experience for all children" in their formative years, which she defined the equity versus equality debate. She explained, "I have been in science and science education since the early 1970s. Never have I been through what everyone else seems to feel is discrimination. My male science teachers were very encouraging when I showed an interest in geology and were very supportive in science fairs, etc. I was never expected to be better than my male counterparts and I can truthfully say less was also never expected from me. I was fortunate when I was growing up that my parents encouraged me in allowing me to take extra science classes during the summer, play in creeks and at my dad's electronic work bench. I had equity with males because I grew up experiencing what males often do. You find some of the same problems in inner-city and suburban children who don't have some or any of the experiences of manipulation of equipment, etc., that some others might. Equity, not equality . . . The idea that we are singling out women is just as wrong as women being forgotten in the first place."

Life in Kindergarten through Twelve

The years between elementary and high school can be the years that separate those that continue in science from those who lose interest. Because many of the building blocks of a science education are set during elementary, junior, and senior high school, a student who is turned off to the field or any of its component subjects early in life is unlikely to catch up. In a standard questionnaire, I asked women from the University of Minnesota to respond to a number of questions. They ranged in age from 21 to 37, representing a variety of science and technological fields as well as racial and ethnic backgrounds. One question sought to determine if respondents' elementary, junior, and senior high school experiences were generally positive or negative.

A typical answer came from Theresa, a 24-year-old doctoral candidate in astrophysics. Her specialty is studying the molecular gas content of distant galaxies. She recalled that "school before college sucked. I liked classes and teachers, so other kids didn't like me, and I was afraid of them because they were 'so cool.' This just made me study harder out of boredom. Junior high was fun because my mom sent me to an alternative school—she was worried that I was too compulsive about classes and had a hard time making friends. Both true. And this school was great."

A common theme among respondents was the general feeling of being out of place among their peers, especially in the upper grades. The influence of peer pressure to *not* excel in difficult classes was cited by several women as a social obstacle. This negative sentiment was tempered to some degree by the inspiration provided by high-caliber teachers, who were not necessarily female. Hillary, 26, is working toward her PhD in physics. She says "I enjoyed science in high school. My physics teacher was a good person who had lots of fun demonstrations. However, it wasn't until I got to grad school that I reflected and noted all through junior and senior high, undergrad and grad school, I *never* had a female science or math instructor."

Teachers have an enormous influence on girls' educational futures, even more so than do peers, according to another American Association of University Women (AAUW) report, *Shortchanging Girls, Shortchanging America*.[5] The importance of role models—individuals who stand out as successful examples for emulation—no doubt plays a big role in girls' career choices and aspirations, and teachers are first and foremost on girls' lists of role models. In fact, of the 3000 girls surveyed in the United States for *Shortchanging Girls, Shortchanging America*, nearly three out of four elementary school students and more than half of the high school girls wanted to pursue teaching as a career.[6]

Curricula and Camps, Programs and Pep Talks

There has been a concerted effort to develop mechanisms to encourage girls and women to enter and stay in the sciences. A part of this effort is mandated or protected by law: Federally funded educational programs, such as those found in public schools, are affected by Title VI of the Civil Rights Act of 1964, which prohibits discrimination based on race, color, or national origin; and Title IX of the Education Amendments of 1972, which prohibits discrimination based on sex.[7] Some programs are geared toward very young girls, others toward high schoolers and undergraduates, and still others toward full-fledged women scientists, working in the field. In this section, I'll provide an overview of methods ranging from curriculum modifications, to social events, to mentoring.

Since the mid 1960s, there has been an awareness that girls and minorities are underrepresented in science classes from secondary schools to colleges.[8] In response, a variety of classroom materials were developed that made it obvious to students that women have a place in the field. One example of a simple, inexpensive, but very striking means to demonstrate this is the Women in Science and Mathematics poster, created by the California-based National Women's History Project. The poster depicts 20 contemporary and historic female scientists and their specialties in a well-designed, full-color format. A student would be unlikely to miss the point if this were hanging on her classroom's bulletin board. Innovative videos and books geared toward school-age children are also available to teachers wishing to illustrate the possibility of women's participation in science and engineering careers.

The K–12 curricula are also evolving to meet girls' needs. The EQUALS program was established at the Lawrence Hall of Science, University of California at Berkeley to train educators and parents in methods to inspire students, particularly females and minorities, to "understand, enjoy, and participate fully in mathematics and computer education."[9] The program has developed workbooks that parents can use at home to do "family math" and an eight-day course that teachers can use in the classroom called Math for Girls, and it provides teacher trainings and workshops throughout the country on mathematics equity and education.

As girls enter adolescence, role models and mentoring become effective. PRAISE for Girls is the Project Roanoke Awareness in Scientific Education program. The purpose of PRAISE, which brought seventh-grade, primarily minority girls to Roanoke College to spend the day with young women science, computer science, and math majors, was not necessarily to supplement the support already accorded to girls enrolled in the gifted and talented program. Project director Jean Carpenter stated it this way: "That isn't really our philosophy. It's more to [encourage] the girl who might not even have her parents look toward those avenues, but has the aptitude and the ability as the teacher might see it in the classroom." PRAISE participants attended college science classes, spent time doing lab activities, and interacted with their mentors, all of which was intended to encourage them to take a full complement of high school science classes, because as Carpenter put it, "If they don't, they can't participate in college at the same level as those who do."

The National Science Foundation betted on advanced technology to advance the cause of science awareness on the part of both girls and boys when it awarded a grant to Dr. Linnea High at Western Illinois University. In 1994, High produced a series of four satellite television programs on women in science that were broadcast to schools nationwide. The programs featured taped segments on historical figures and live, interactive interviews with working women scientists who responded from the studio to questions phoned in by students. The programs were taped and are now available to schools along with accompanying classroom materials.[10]

Science support strategies are also useful in college, where women continue to leak out of the science pipeline. Several years ago, Dartmouth administrators noticed that the 45% of incoming freshmen women who had indicated an interest in science had somehow shrunk to 15% actually majoring in science. Officials set up a mentoring program called Women in Science (WIS) to plug the leak. Participation in WIS has grown steadily as undergraduate women hear of its benefits, particularly its paid scientific research internships with faculty.[11]

Some of my survey respondents spoke of attending special science events created for women and girls, such as weekly support groups, seminars, and summer science camps. While many valued their existence, participants were not uniformly positive in their evaluation of these programs.

Anna, aged 25, mentioned that she was "very active in several women in science and also women in leadership programs, specifically because part of the reason I find the field of physics interesting is because it has so many unsolved sociological as well as academic problems. There are several support networks here at the university, and I've tried to support them in return to the best of my ability."

Theresa, another grad student, was reluctant to provide full approval: "I don't really agree with any of these things except for *maybe* support groups. The rest, I think, simply tend to man- and establishment-bash and cause our male colleagues to be threatened and therefore further isolate us. Support groups can also do this, but are more variable."

Hillary has been active as a participant and a mentor: "I regularly attend the pizza lunches for women grad students at the Institute of Technology. I also attend the brown bag lunches for women in physics—both are to create a better environment for women. I helped with a Girl Scout event to introduce girls to science. I also went to small colleges and talked about life as a grad student."

Beyond the clubs and camps for younger women are the programs offered by foundations and other institutions that aim to assist women's professional development. The Association for Women in Science (AWIS) is a national organization that provides a support network for women scientists across the country, leadership development programs, and a voice in Washington, DC, on issues affecting women in science. Other examples are the "Women at Lotus" employee group; a National Institutes of Health program that funded female biomedical researchers returning to science after taking leave for their families or children; and the many National Science Foundation (NSF) programs aimed to support women's progress in the sciences.

Naomi, an associate professor of chemistry, described her experience as a National Science Foundation Distinguished Visiting Professor: "That the NSF cares about women is truly remarkable to begin with! The experience taught me many things, one of which is that there are truly very few women in higher education in science. I can practically name each and every woman I met at three universities in their science and math departments! I was astonished!"

Scientific Higher Education: The Apprenticeship Explained

Undergraduate Programs. The normal undergraduate science curriculum requires early specialization and a variety of courses in support of the major field, which tends to limited choices for nonmajor classes. The most anointed faculty (at least at large universities) are those with substantial research programs and accolades—they also tend to be less interested in teaching vis-à-vis research. Institutions encourage this focus on research, because along with scientific research grants come considerable administrative and overhead budget add-ons. For example, faculty members writing up research grant requests at the Massachusetts Institute of Technology (MIT) must add an additional 43% of the total requested amount to cover indirect costs, such as physical plant operations, maintenance, libraries, and general administration. In fiscal year 1990, MIT took in $292 million of federally funded research grants and contracts (out of total budget revenues of $1.06 billion).[12] In 1990, it was revealed that overhead charges at Stanford University for programs through the Office of Naval Research had reached a whopping 78%. Financial auditors trailed "overhead" money that had been spent on a Stanford yacht and university parties.[13] Numbers like these show why research monies figure so prominently in the economies of most universities. It is open to debate whether the recent clamoring by liberal arts colleges to develop "undergraduate research programs" is an honest attempt to enrich the science preparatory experience for students or a barely disguised move to garner cash for their bottom lines. The fact that institutions must incorporate scientist training into their research in order to win grants has fueled the overgrown pool of scientists having difficulty finding work in their fields, which is an enormously important issue in science today that is addressed in more depth later in this chapter.

The largest block of most students' time in science courses is spent in introductory laboratories, where the hands-on principles of the discipline are frequently taught by graduate students scarcely older than the students themselves. Classes are almost invariably conducted on a lecture, not discussion, basis. Most teachers do not attempt to encourage an exchange of ideas. This educational style reinforces science as a body of knowledge and tends to underplay its other important aspects, among them scientific research as a creative process, science as a product of culture, and the role of creativity and intuition in scientific discovery. Sexism is also present in the classroom: A mechanical engineering undergrad posted to the Internet the resistor color codes mnemonic provided by her male professor: Bad Boys Rape Our Young Girls But Violet Gives Willingly.

According to the 1992 NSF Report on Women and Minorities in Science and Engineering,[14] those female freshman who express an interest in a science major are, like their male couterparts, better prepared academically than the class at large. Of students expecting to major in a science or engineering field, 39% of the females and 38% of the males reported a high school grade point average in the A range. However, the genders differentiate markedly on their choice of eventual career. Female students' choices run to clinical psychology, law, and social work fields, while almost 40% of the male students majoring in a science/engineering field intend to pursue a career in one of the engineering disciplines.

In 1989, US institutions granted almost 308,000 bachelor degrees in science and engineering. About 40% of these degrees were earned by women,

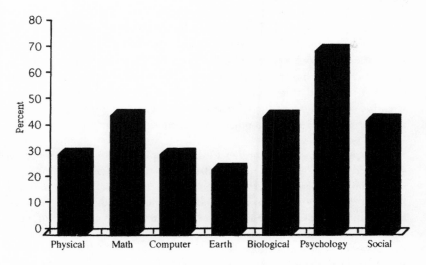

Figure 1 Δ Women's share (expressed as a percentage) of all bachelor's degrees in natural social sciences awarded in 1989. SOURCE: Patricia E. White, *Women and Minorities in Science and Engineering: An Update* (Washington, DC: National Science Foundation, 1992), p. 23. Reprinted with permission.

up from 35% in 1979. Figures from the NSF report[14] indicate that the social sciences, including psychology, account for almost 60% of the science bachelors granted to women in 1989. The next most popular science major was biological sciences (including agricultural science). Attracting relatively few women were mathematics (6.3%), physical sciences (3.8%), and earth sciences (including geology, atmospheric, and oceanic sciences), at less than 1% of graduates. Overall, women represented 47.2% of all undergraduate science degrees granted in 1989. A breakdown of women's shares of various science majors is given in Fig. 1.

The remarkably positive reports coming from women who have attended all-women schools and colleges say much for gender-segregated education. In a letter to the editor of the *New York Times* on August 17, 1993, Whitney Ransome and Meg Milne, executive directors of the National Coalition of Girls' Schools, pointed out that students at all-girl schools "take math and science classes at double the national average and do well in physics." After enduring the patronizing attitudes of her male peers in high school honors classes, Gina's experience at Wellesley salvaged her sinking hopes of a scientific career. "Everything I did for 4 years was centered on women's advancement," she reports. "I cannot possibly express with a thousand words how much attending Wellesley helped me affirm my decision to pursue a scientific career. I can't emphasize enough how important it is for women to take part in women-centered programs. With all the sexism still in the world and in the scientific community, we should take advantage of every opportunity we have to reduce it and to affirm to ourselves that we do belong."

Graduate School. According to data compiled by the NSF,[14] women comprised about one third of the 401,569 graduate students enrolled in science and engineering programs in 1990, up slightly from the 31% recorded in 1982. The number of women enrolled in science graduate programs increased by 30% over 1982 figure (compared to a 12% increase for male students). At the masters degree level, women represented 41.9% of all science degrees granted; at the doctorate level their share slips to 33.1%. As shown in Fig. 2, women's share of science degrees vis-à-vis men drop at each step of the educational ladder.

For most fields of science and engineering, a postbachelor's degree is considered an entry level requirement. This is of course true for anyone who aspires to teach on a university level. The goal of graduate school in the sciences, as with other fields, is to produce, through a combination of course work and practical experience, a fully functioning professional. In the sciences, however, this has taken on new meaning. Graduate students perform research to complete their doctoral programs and to provide grist for their thesis/dissertation. Professors for their part are expected to have healthy research programs, staffed by graduate students and paid for with predominantly federal grants. New professors' progress toward tenure is marked in part by the success of their research efforts, so it is essential to find and employ the top-performing grad students. The educational experience at most graduate institutions is one where the research enterprise is taken extremely seriously by all concerned. Many grad students may also support themselves by holding teaching assistantships, wherein they teach introductory lab courses to undergraduates. This role is usually played down, however, and is unofficially considered an impediment to beginning true graduate work (i.e., a research project).

Although varying from department to department and lab to lab, the premium placed on publishable results brings about a work ethic in most

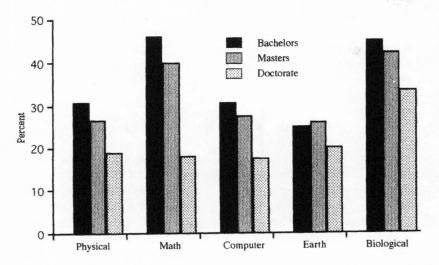

Figure 2 Δ Women's share (expressed as a percentage) of all bachelor's, master's, and doctoral degrees in selected scientific fields awarded in 1989. SOURCE: Patricia E. White, *Women and Minorities in Science and Engineering: An Update* (Washington, DC: National Science Foundation, 1992), p. 24. Reprinted with permission.

research groups that places a great deal of value on long hours and dedication. This ethic probably suits best the unmarried/unattached students or those with few interests outside the laboratory; students with families or with strong connections to an outside community (social, political, etc.) often find themselves the subjects of queries and pep talks from their advisor. This persistent work ethic follows a scientist throughout her career, and is addressed in depth in Chapter 4.

The Postdoctoral Experience. In years past, the normal course for a young scientist after receiving his or her PhD was to obtain permanent employment, either in industry or in academia. For a variety of reasons, which will be described in more detail at the end of this chapter, these permanent jobs are increasingly difficult to find in the sciences. More and more newly minted PhDs in scientific fields are availing themselves of the postdoctoral fellowship, or "postdoc," essentially an extension of the apprenticeship begun in graduate school.[15] Designed to better prepare a scientist for a research career, postdoctoral appointments place the young scientist in a more active research analysis and management role. Postdoctoral researchers are supported by the research grants written and won by the principal investigator (usually, the professor) on a term basis. Hence, these positions are temporary and paid for by "soft" money, providing a paycheck and experience but no job security.

Along the Way: The Social Influence

Peer pressure is a definite factor in determining where young women's scholarly paths take them. Adolescents are squeezed to conform to the norms of their peer group, norms that do not typically celebrate academic achievement. Another area that is gaining in importance for its influence is the effect of family pressures and attitudes on girls' career choices. Researchers Linda Trigg and Daniel Perlman looked at women entering high-status, "nontraditional" careers and found in one study that certain social factors, including, interestingly enough, their mothers' attitudes (not so much their fathers' or significant others') were important factors in the students' decisions.[16] It's instructive to hear from women regarding questions of social and family pressures. Did they experience them in their formative years? From whom? And how did that affect their decision making?

If the combined experiences of my respondents can paint an impression of how girls interested in science are treated by their peers, then we've nothing to be proud of within US schools. All the cultural convictions regarding appropriate roles for women are brought to bear on the girls who might consider nontraditional paths. The people who are closest to girls overtly or unintentionally convey their opinions about women's roles, leading girls who may have interests outside the norm to question their self-image, interests, and career goals.

Now a medical researcher, Julie had a mother who pecked at her budding interest in medicine, sneering: "So you think you know everything, don't you?" and "You think you know more than the doctors?" Her nearest sister called her a "nerd." Her schoolmates often ridiculed her for her academic achievements. Yet now, as an adult and a successful scientist, Julie reports that her family has given her much more support. This change of heart illustrates the importance of firsthand experience with women in nontraditional roles. Once we encounter women who have broken through our prejudicial notions of what women can or cannot do, it appears that we can begin to alter our own perceptions. Certainly it's not surprising to come across a female attorney, physician, or stockbroker today, unlike 50 years ago when it would have been rare.

It doesn't seem to matter if only one person in a woman scientist's personal life is critical of her career choice—this sole voice can inflict plenty of damage. As Hillary, a bright, midwestern PhD candidate in physics said, "Everyone in my life has been very encouraging to me, except my mother-in-law, who can't figure out why I'm doing all of this since I'll eventually stay home with the children." Hillary had engaged her mother-in-law in rational discussions as to why she was pursuing an academic career, but to no avail. Despite the overwhelmingly positive messages she was receiving from her husband, her parents, her siblings, and her friends, her mother-in-law's disapproval was a throbbing thorn in her side. Many of the women I corresponded with mentioned an experience like this, where a single naysayer had made them second-guess their career decisions at one time or another.

Family aside, quite a few respondents mentioned men's and boy's reactions as factors in their academic pursuits. From Theresa I heard that, "Boys said they didn't like me because I used 'big words.'" This reference to intelligence being an unattractive trait in girls was repeated quite a few times. Because of her brains, Gwen endured a conspicuous lack of boyfriends, because, as she states, "I was, of course, not the typical girl in high school, since I didn't hide my intelligence like many of the other girls, and I didn't act stupid or giggle around the boys to make them feel smarter. So, even though I was very popular and was involved with everything, the boys stayed away. I really didn't care what they thought and knew that I had much more ahead of me."

Girls, too, are guilty of barraging their academically gifted peers with negative messages and even insults. The feeling of being slightly out of place applied to Marie, who explained, "Girls thought I was unusual but funny, a bit distant. Boys thought I was uninteresting (especially under the age of 18). I did spend quite a bit on energy in high school trying to figure out what was 'wrong' with me—it was almost as if I had missed a meeting, missed the memo that said, 'Girls should now abandon math and science classes and become interested in soap operas.' I really didn't understand many of my peers. Despite my participation in school sports and other activities, I felt isolated."

Sometimes girls interested in science have the last laugh at their cruel peers. Carol responded to this question from the Internet, and wrote about a pretty typical teenage social scene. She described her outcast status: "I grew up in a small Texas town, and my high school graduation class had only about 60 people in it. Throughout junior high and high school, my classmates shunned me as a 'brain.' They were interested only in how the football and basketball teams were doing, who was dating who (and who was sleeping with who), and where to get their next bottle of wine or bag of grass. Since I was more interested in my academic pursuits, I didn't fit in and was harassed as a result. It wasn't until the spring of my senior year that the situation changed appreciably. With graduation looming, it finally dawned on them that without a college education, which they had not prepared for, they were going to be stuck in low-paying dead-end jobs. At that point, a small but discernable level of respect emerged, and some of them actually made an effort to be friendly."

After hearing it so many times during the course of this project, I began to wonder what words students interested in the sciences would use to describe themselves if "nerd" was expunged from the language. Beth, a Swiss physicist who studied and works in the United States, furnished it for me: "I was sort of a freak." She goes on to say that in her youth she was "admired but not understood. In college there was disbelief of my sincerity of pursuing a career in science, and hostility when my male colleagues found out that I was not in college to get married to a prospective PhD. Hostility came because of fear of competition. Their attitudes made me mad, but never got me into thinking of leaving science. What really helped was that I was not the only woman in my class, had real backing and support from my family, especially my mother, and was influenced by feminism."

Can anything good come of negative peer attitudes toward intelligent girls? Consider Carol's unsettling point: "I think that having to learn as a teenager how to cope with daily hostility prepared me to cope with the hostility I have faced as a female technical professional."

Of course, the majority of respondents were representing the US experience of social pressure in their responses. Connie, a 33-year-old Canadian computer scientist, wrote, "I did not experience any of the following until my current job, and I am sure that they don't do any of these to a male colleague." She then lists: "They told me that I must be a new co-op college student. They told me that I must be a new secretary. They winked at me to show that they are friendly. They told me that women were hired because of affirmative action. They told me that putting some women in offices makes the workplace not as boring. They told me to write down/document their ideas in a group discussion. I saw many times, when a group needed to reach a decision, they only checked with male colleagues at the table. I saw many times, when a female colleague suggested an idea, they totally ignored her. Nobody said 'Yes, it is a good idea,' or 'No, that won't work because . . .' They simply went on talking about something else."

It was also fascinating to hear from a talented astrophysicist who completed her education in India: "The women's college where I got my bachelor's degree was in India, and the teachers (of equitable gender ratios even in the sciences) would nominally be classed as 'traditional' or 'conventional.' However, there was no bias whatsoever against women studying science— only encouragement. The problem of girls in the United States being discouraged from science was a new and shocking one to me. The inequity came at a higher level in my experience; when it was a choice between a daughter and a son going for higher-level studies that required financial resources that were scarce, then the sons got the preference. My master's and PhD degrees were obtained at much 'higher-grade' institutions. Here I experienced a different ambience. My peers there thought themselves to be more Westernized and 'progressive.' Ironically, I noticed that while I did not face any discrimination in admission procedures and so forth, which tended to involve more senior people, most all male, and supposedly of more 'traditional' attitudes, there was often a subtle hostility from peers of my own level. I would often see how women had to work harder to prove themselves. Flaws in males would be attributed to the individual, while flaws in the women were attributed to the fact that they were women. Women that played 'dependent,' naive, helpless, and physically weak, were 'popular' and did not seem to be viewed as a threat. Those that played independent, knowledgeable, and nonchalant were viewed with hostility in subtle ways. While I did not particularly want to be 'popular,' I would have liked to be able to be normally friendly and affectionate to my peers to the extent that I am naturally wont to be."

In a paper titled "The Impact of Family, Peers, and Educational Personnel upon Career Decision Making," Betsy Bosak Houser and Chris Garvey,[17] then at the UCLA Department of Psychiatry and Biobehavioral Sciences, looked at social influences on women who were considering entering trade and industrial vocational programs—clearly fields with few women. The researchers found that women who pursued nontraditional careers consistently received more personal support and encouragement to enter male-dominated careers than women who had entered female-dominated careers. Among those women who had considered a nontraditional field but who ended up avoiding that path, the only significant difference in support they received was from the men in their lives.

It became apparent during my contact with women that it's not unusual for some women's entire self-concept—including femininity and even sexuality —to be bound up with their career choice, and that peer groups and families can be adept at laying on their own assumptions on this matter. Theresa, a grad student in physics at a Big Ten University, goes so far as to say that her siblings feel that she is not "normal," and she told me that she is convinced her family believes she is a lesbian simply because she is a scientist.

Career-based sex-role stereotyping by peers and family is certainly not limited to women. Charles, a physicist turned biologist, relayed a telling anecdote regarding societal pressures and images of scientists. "I was [dating] a female PhD student while I was doing my PhD in another department. When I first met her mother, she took me aside and had a little conversation with me. In it, she explained that women need a man to look up to, and men need a woman who looks up to them. Since I was getting a PhD, as was her daughter, this would not be possible. However, [her mother explained] things would be very different if I was getting an MD. That was the only sort of man who could be good enough for her daughter."

The Two-Body Problem: Married Scientists

The futures that young women see for themselves in science often collide head-on with their expectations of family structure. Of course, not all women scientists expect to marry, not all want to raise children, and certainly some will end up in nontraditional relationships where their science career will take precedence over the career of their partner. But a good many will end up marrying, having children, and confronting the reality that a career in science requires enormous mobility.

Career specialization often leads to a decrease in employability in any given geographic location. As a result, PhD-level scientists must be ready to locate to any city in the country to begin their careers, and must be ready to relocate at 2- to 3-year intervals until a permanent position is at hand. This arrangement is hard enough for a single person; for two married scientists, it is a major career obstacle. Finding two good jobs in one place has taxed many a marriage that was born of long hours spent together in the same graduate research lab. Upon graduation, whose job offer takes precedence? A very high percentage of married women scientists have spouses who are also scientists or engineers; the figures are 80% for female mathematicians and 69% for women physicists.[18] In spite of the ubiquity of dual career couples, the practical problems posed by both women and men scientists have only recently begun to receive serious attention by department heads and administrators.[19] Solutions such as creating tenure-track job-shared positions, offering career placement assistance to the trailing spouse, and seeking soft money start-up funds for the spouse's specialty are all being considered.

Many of my respondents mentioned this type of family pressure as an influence on their decision to go into the sciences. One respondent saw that "the greatest potential barrier to my success would be marriage and family, effectively decreasing my flexibility and mobility and time commitment. Marriage versus career is a real conflict. Very few men will sacrifice their career goals so that their wives would benefit. Otherwise, both partners usually must make some career sacrifices." Thea, a grad student in astronomy, is married to another astronomer. She reports that "My advisor has said that HE didn't get married until he was out of grad school, implying that I can't afford to take time for my husband. Others not in science, by the way, are saying the same things are being done to young people in their fields. It's supply and demand, and entry level jobs are getting scarce."

Women Scientists in Popular Culture

In 1957, two anthropologists, Margaret Mead and Rhoda Metraux, surveyed schoolchildren on their ideas about scientists. Boys were asked what kind of scientist they could imagine themselves becoming. Girls addressed themselves to what kind of scientist they wanted to marry.[20]

For many of us growing up in the 1950s and 1960s, our only imaginable occupations were limited to those we experienced in our immediate surroundings: parents' employment, teachers' interests, siblings' and other relatives' careers, and what we knew about our neighbors all contributed to what we could concretely grasp as life's possibilities. Television, books, and movies provided the bulk of the rest of those possibilities; their images of fiction and reality brought much of the rest of the world into our living rooms and reading chairs. Because of the importance of these images, which have become so pervasive in the last couple of decades, I asked my respondents to describe any popular media images of women scientists that they recalled from their childhoods and to describe the images they encounter now as adults. Their responses give us some idea about how difficult it might be for a girl to imagine herself as a scientist if her immediate environment did not portray science as a living, breathing example of a career option. Who knows what today's youngsters images' of scientists are, given that respected TV science programs such as PBS's "Nova" attract a relatively small percentage of audience share and that well-intentioned programs for younger audiences, including "Newton's Apple," Disney's "Bill Nye the Science Guy," and CBS's "Beakman's World," though they feature girls and minorities in supporting roles, are hosted, predictably enough, by men—and white men at that.

In responses received via the Internet from women and men scientists, I learned that the prevailing media view of women scientists involves low-power, sexually unappealing, bespectacled lab drones. A 21-year-old physiology major was very specific: "Popular images of women scientists in the media are primarily unattractive, with some form of disfigurement; unmarried (unless it is to their work); usually subordinates (technicians, etc.), never the main decision maker; emotional, or on the other end of the spectrum they are 'cold.' "

A molecular biologist states that, "For the most part, scientists are, independent of gender, depicted as nerds who wear lab coats. The Combat roach control commercials with their scientists in lab coats are more the norm: 'We're nerds and we think of nothing but bugs!' "

It is not just women scientists who are portrayed in an unflattering light. A young woman in her second year of grad school in astronomy said, "Usually men are asexual geeks, which is as upsetting to me as the way women are portrayed. Generally, it is a pretty pathetic representation and not realistic in any way."

We've already looked at the unpleasant social and family environments faced by some girls who pursue science interests during adolescence. Obviously the negative images of women scientists bandied about in high schools get a wider audience when they are disseminated through TV and film. What gives here? Did prime-time screenwriters all attend the same high school? Or did all the creeps in high school see the same TV show? Just where did the negative images of scientists come from?

A good guess is that they derive from the fact that science is difficult. Not everyone can follow the increasingly complex mathematical concepts that form the foundation of the physical sciences. Brains function differently. Some people can accurately identify a Georgia O'Keefe painting at 20 yards, but are unable to solve for x; others do extra-credit word problems for fun, but could not write a coherent essay to save their lives. The truly gifted can do both, and quite often go on to become phenoms in their fields. It's no secret that humans can be a petty, jealous lot; and those who have an easier time of it can become the targets of those who do not. Children who breeze by their classmates in schoolwork are prime targets for taunts and bad feelings. Kids and adults who are enamored of their work at the expense of developing friendships or other interests are branded as oddities. Not all bright kids fit into this description, of course, and go on to become homecoming royalty or sports stars. Still, scientists are dealing with a public that functions on the lowest common denominator, and to the person on the street, what scientists do, and the manner in which they do it, are outside the norm.

Thus the messages that popular culture broadcasts about scientists reflect a decidedly low-brow, anti-intellectual bent, resembling the crude insults of a teenager more than the considered thoughts of a mature adult. Pop culture takes us all for fools, as described by Andrew, a scientist working in an industrial lab: "Female scientists in the entertainment media are almost invariably 'babes'—an archetypal example being the Kelly McGillis character in 'Top Gun' who was supposed to have a 'PhD in astrophysics'! [This is a] good example of how terribly confused Hollywood is when it comes to science in general, as if astrophysics was intimately connected with airplanes! What do they [women scientists] do? We don't know. The movies are quite uninterested in showing us that, not to mention incompetent to do so (this goes for male scientists too, pretty much). What do they wear? Often, quite unattractive and unstylish outfits, until they have their transforming encounter with the unscientific male protagonist."

As John pointed out, there is a flip side to the image of female scientists being unattractive or unapproachable, but it carries the imagery to the polar opposite. The "scientist as sex object" theme was picked up by a 24-year-old female grad student in space physics, who said, "I never took note of any images of women scientists—I guess that's fortunate. Now I get very offended very often: are they ever wearing anything but high heels under their lab coats? And what happens to TV women scientists over the age of 27? Are they retired?"

The objectification continues. An outspoken young astrophysicist says, "The image of the 'oh-so-sexy scientist by day, all-bosom-heaving woman by night' is ubiquitous, offensive, ridiculous, and unfortunately, powerful. The message is that it's only OK for me to be smart if I'm damn beautiful as well. I bought into this for a long time, and really felt, as late as age 21 or so, that I'd trade my brain for a supermodel's body any day."

Perhaps it is preferable to be portrayed as sexy or nerdy than as insane, but popular media will apparently give us that image too. An astronomer working at the Space Telescope Institute shares her view, "It seems that scientists in the movies are often depicted fairly generically, wearing white coats—even the astronomers, which is a big laugh—and occasionally as absent-minded, crazy, or megalomaniacal."

The popular media is not alone in broadcasting sex-role messages. A grad student in medical research referred to a respected journal in her specialty when analyzing its treatment of women: "Women are rarely in the lead. Most often photos show women being 'talked to' by men. The journal *Cancer Research* has cover photos of scientists, almost always male! From January to December 1993, there were 70 men on the cover and only 14 women. Of the 14 women pictured, 10 were staff only working for the men pictured. Also, while men were pictured alone on 17 covers, women were alone on only two covers in 1993."

Though he tried to think of even one, this sorry state of affairs was wrapped up by Peng, a physicist at a midwestern university, who laments "Strangely enough, I cannot recall any positive images [of women scientists] from the media. Not at all."

Role Models

Many women mentioned the profound influence of role models on their decision to enter science. Interestingly, the role models that they admired were not necessarily women. It's fortunate that women can appreciate male scientific role models, since there are indeed so few women to emulate. One respondent even mentioned an important media role model that influenced her decision: "I am somewhat embarrassed to admit that Spock from 'Star Trek' was a role model. He was the first truly 'cool' scientist in pop culture, who was not only always right but who was absurdly conservative in his statements (his stated probability for being incorrect was infinitely smaller than his batting average) and who was a superb physical specimen besides. Ah, those were my values in junior high."

As she grew up, Spock faded into the background for this respondent, and was replaced by the "many women I regard as role models, in the sense that they have made it, that they're obvious examples that someone can be a woman, do woman-things like give birth, and still have wonderful lives as astronomers. These women range from graduate students to senior faculty. Sometimes I wish there were more of them, when I'm feeling discouraged. It can be easy to dismiss a few examples by saying, 'Well, they're super-scientists.' "

Vicky, a 26-year-old grad student in astronomy, also had a succession of role models: "I had a sixth grade teacher who allowed me to write a term paper on the whole solar system, even though she suggested the topic was too broad. I got an 'A.' This particular teacher also had us build model rockets and launch them. In the beginning of my planetarium training, one man in particular was very inspiring to me. He joked that someday I could say he taught me everything I once knew. [I] harbor hope that I may someday advance to a position similar to the one he holds, but I feel that he had an advantage that I don't. He had a wife who could move around to follow his career. My husband has established his own business as a carpenter, and since he will always make more money than I will in astronomy, we will probably choose not to move around to advance my career."

Unfortunately, Vicky has not found any female role models in her grad school career. "There is not *one* woman faculty member in the astronomy department [at this university]. There is one woman who will probably be joining the faculty next year, and if she gets tenure, she will be the first woman in the 30-year history of the department to do so. There is not one woman here to be a role model or mentor. It's men or nothing."

Accepting a male figure as a role model or finding no role model at all is a definite possibility for women scientists. I heard again and again about positive experiences with men. Thea, who began her science career later in life, said, "I didn't have any role models as a child or teenager or young adult, and I suppose that is one reason that I didn't do science until I was 30. The person who gave me my first part-time job in astronomy, the department head at the time, was and is a role model for me. He says, 'I don't know' a lot. He thought it was just great that I was doing science as a reentry—the first person I had met who didn't think it was weird."

Women who connected with female role models were truly enriched, as was Sonia, recently graduated with a BS in physiology: "As a student, my role model was a high school biology teacher who was a woman. It's because of her that I went into the health sciences rather than computers or engineering. What triggered this was her extreme enthusiasm to learn rather than memorize. She encouraged research in the sense that if you weren't satisfied with an answer that someone gave you, then [you should] do something about it! Also, she wasn't a single-minded teacher focused only on her topic. We discussed things ranging from souls, to science breakthroughs; she took us to see an autopsy, we stayed after school to dissect sharks (our first experience of that kind), she even let me do a full class presentation on the brain just because I wanted to. She was wonderful! She was also a mother with a daughter who was a bit younger than us; so she would tell us her experiences with her daughter and would learn from us as well. She also encouraged the girls in the class more, since she herself had faced difficulties in wanting to become a scientist in her time."

Lisa listed so many role models in her family milieu that her career choice seemed predestined: "Mother: epidemiologist; Grandfather: orthopedic surgeon and naturalist; Grandmother: anthropologist and naturalist; Father: physicist; Brother: geologist . . ."

Memories of supportive family members rang true with Yvette, a 23-year-old clinical engineer whose father instilled in her an early and abiding interest in science: "My dad [was an] electrical engineer at NASA. I can best depict our relationship with a memory of him explaining to me how the distance of the horizon could be determined as we looked out at the gulf from Galveston Beach—he began deriving equations in the sand with a stick. I was only 5."

In a dismal turn, several respondents noted that even though they had contact with established women scientists, these women were either reluctant or poor role models. Charlotte, a visiting scholar in physics and astronomy at a prestigious university, said "Unfortunately, I have never had a female scientist as a role model, despite the fact that I've known quite a few of them. This includes those I've seen in the media and those I know personally. Of the women scientists I personally know, most that are 'accomplished' are bitter and unfriendly to me and other women. I have never had a good relationship with an accomplished woman scientist that caused me to consider that person as either inspirational or a role model. This has been both confusing and irrational to me."

Charlotte's bad experience is echoed by Janet, a biochemistry grad student who finds that, "Older women in science seem to be less inclined to give younger women a hand, I guess because they had to fight so hard to get where they are." Continuing, she makes the observation that, "younger female scientists seem more sympathetic."

Good role models and mentors are sometimes molded by their own life situations, agreed Penny, who observed, "In my experience it is the 'old guard' who have daughters that are the most comfortable with women in science. It is as if they are willing to extend the opportunities they desire for their daughters to other female students. The hardest segments to [encourage to support women] are recently tenured professors who have small children and very dedicated wives—they don't see why we have anything besides science to do—and secondly, women scientists [who] 'toughed it out' and expect all women to follow their path. Many younger (18–25) folks in science appear to have never considered gender roles in science, they have always been trained together. I was the only woman in many of my undergraduate science classes. This made me feel that women were not interested in science. I was very fortunate to attend graduate school where the doctoral students were mostly talented women, which was fundamental to changing my attitude about women in science."

Irena, a physics grad student, feels that she owes it to future students to be a female role model herself: "I feel it is my responsibility to be a role model for future women scientists. Because there are not many physics professors who are women, I'd like to become one. Eventually, I'd like to hold some sort of decision-making position, like a dean of something, to help make decisions that include women's interests."

Irena may find that becoming a role model will be easier than becoming a mentor, because many young women scientists in academia are pushed to the limit timewise. I had a conversation about this with Elaine Seymour, the Director of Ethnography and Assessment Research at the University of Colorado's Bureau of Sociological Research. Dr. Seymour, who has studied the problem of high attrition rates for female undergraduates in science, pointed out that the women in the best positions to mentor students are often the most pressed by their institutions. She said, "They're overworked, they're strapped, they're on every damned committee in the university that wants women as a token, they're probably dragged into some women's program where they have extra responsibilities, and they're still trying to compete with men on their own terms. It's a bit hard, you know?" She stressed that one of the reasons that older female scientists may allow younger ones no concessions is not out of spite, but because they feel that young women are "going to have to tough it out too—it's not going to help them to soft-soap them."[21]

I received a charming but ultimately troubling response from a male assistant professor of physics, who felt that it was a female role model that encouraged him to enter the field. English is not this respondent's first language. According to him, "In the middle school, I had a beautiful girl classmate who was always doing better than I did in everything, including courses in science. I dreamed to be as good. This [was] the first trigger for me to study hard in school. In college as well as graduate schools, I met many great women students, who performed great in science and engineering. But strangely enough, when I started to work, I no longer have the same chance to see bright women scientists. I heard names, but seldom meet anyone personally. Why? And what has happened to those great many women students?"

Affirmative Action: A Double-Edged Toehold

The efforts to bring more women into the sciences also include top-down measures, such as requirements of gender balance on funding proposals and affirmative action/equal opportunity hiring policies. Mary E. Clutter, the assistant director for biological sciences at the National Science Foundation, made waves in 1989 when she established an "antisexism" policy in her directorate that required grant applicants to include women presenters on conference agendas.[22] Clutter saw plenty of top-notch women researchers who were being looked over for speaking engagements and seized an opportunity to increase women's visibility and participation by revising the way grants were handed out. Her action broke down the male-dominated power structure with a sledgehammer, allowing women to take their rightful places on the podium alongside their male colleagues. This kind of direct action annoyed some, as shown in an editorial in *Nature*, where a writer lamented, "There is no evidence that sex is related to success in scientific research, and no inherent justification for holding women out for special treatment as part of a formal policy carrying the bludgeon of budgets."[23]

Affirmative action measures are tremendously controversial in the science establishment. I discussed the topic with a group of male scientists I know, all of them white and all from comfortable but certainly not plush backgrounds. They tried to convince me that gender-based affirmative action is codified discrimination—no more, no less—and that it should be recognized as such and jettisoned in favor of some other, "fairer" way of assuring that women are not discriminated against in hiring practices. "Fairer" methods that they suggested included masking the names (and thereby, supposedly, the genders) of job applicants from their letters and resumes, and making sure that women are on hiring committees (even though many departments would have to look elsewhere to find a female peer to sit on the committee, and how could an outsider really sway the thinking of an established group). Both of these suggestions have an odd logic; and in any case, they don't particularly address the purposes of affirmative action, which really do include reaching out to get more women into the loop.

What are the outcomes of affirmative action? There are a couple of ways to look at it: affirmative action in one sense assures that the best candidate, *even if she is a woman,* is allowed access to a job. Affirmative action in this sense opens up employment in groups and departments that might never have allowed women even the opportunity to join. Female-free scientific groups are alarmingly common in the United States, and one gets the impression that men working in these groups are quite comfortable with the status quo. A second outcome of affirmative action, which in particular draws the ire of men, assures that a given job, typically in a department that has no or few women on staff, will be filled by *a woman,* no matter what comparative qualifications male applicants bring to the process. I have heard from scientists who believed that in order for universities to comply with affirmative action goals, they have slated many of the few academic positions now opening up specifically for women. Some men, even though eminently qualified, felt that they might as well give up and skip the application process, because they had already been precluded from selection by their sex. These men may have a point: In keeping with government affirmative action regulations, institutions must make every effort to make up for any statistical deficiencies of individuals in their workforce by sex and race. Of course, this leads to some positions being effectively closed to men, or to whites, or to both.

The concept of reverse discrimination would probably never have been mentioned by my friends in a better job market. Long discussions of it appear regularly on electronic discussion boards like the Young Scientists Network, where disbelief, anger, and disgust show clearly through the words of young scientists. There is little reason to believe that these men are woman-haters; they take pains to declare that all discrimination, even the "reverse discrimination" of affirmative action, is wrong, and they endlessly brainstorm other mechanisms to increase women's scientific participation in what they believe to be fairer ways. These men appear to be more hurt than on a jihad. What's troubling about their posts is the amount of blame that they place on affirmative action for their personal job-hunting quandaries. In hearing so much about the evils of affirmative action, one might begin to think that the women who attained positions through it were unqualified; that women in droves were snatching the only desirable scientific jobs; and that affirmative action was a plot, specifically designed to keep earnest, talented, and promising young male scientists out of work. Over and over again goes the refrain, "I am not responsible for keeping women out. Why must I now suffer for the sins of my [male] forebears?" In some ways the harping on affirmative action becomes disturbing. It's not difficult to make comparisons between this subculture of male scientists and the invidious attitudes of other historically powerful groups, down on their luck and seeking someone to blame.

It is hard to counter the idea that affirmative action is reverse discrimination, and even harder to convince a perennially underemployed, overtrained 35-year-old male that he is part of a privileged group. In her book, *Transforming Knowledge*, author Elizabeth Kamarck Minnich[24] puts her spin on the issue:

> Provisions for the continuing exclusion and devaluation of those defined and treated as inferiors are properly called "racist," "sexist." Provisions for the *disruption* of such systems are, on the contrary, antiracist, antisexist. There is no reversibility here. There is, instead, recognition that that which has oppressed people must be combated in terms that openly and explicitly take that oppression into account.

The affirmative action question is a tough one, even for women who have benefited from it. Sometimes especially women who have benefited from it, who have been subject to blatant slurs and ongoing professional speculation, such as Tamara, who noted, "I suspect my assistant professor position was available only for women or minorities. This caused a certain amount of friction with my peers initially until I could otherwise prove myself. Also, it caused somewhat of an imbalance within departmental research areas." I interviewed women who were told by their male colleagues that the only reason they'd been hired, won an award, been asked to present a paper, and so on was because of official pressure. Their legitimate talents and accomplishments were discounted by their detractors' near-obsession with affirmative action. As Sharon, a computer scientist with NASA, experienced: "The problem I have with affirmative action is that those who don't benefit from it believe the only reason I am where I am is because I am a woman. I have to then prove my competence instead of having it taken for granted."

From the amount of tension my limited survey could pick up about the issue, it's obvious that affirmative action programs need to be carefully monitored and scrupulously applied, explained, and most of all evaluated. As one woman, a vice president of engineering in a software company, put it, "If there is ever the glimmer of a possibility that I am being offered something (job, recognition) simply because of my sex, I will turn it down—I want to be the best."

Even though she resents the affirmative action backlash, Carol nonetheless supports its continued application: "If anything, the backlash from white males against such programs has caused problems that I do not think I would have otherwise faced. Nevertheless, I support affirmative action and EOE, primarily because it's a good way to keep people churned up and talking about the issues, but also because it provides a means for pursuing legal redress in the most outrageous (and well-documented) cases."

All Dressed Up in Lab Coats and Nowhere To Go: Today's Job Market for Scientists

In the mid 1980s, an analyst at the National Science Foundation drew up a draft report on expected trends in the production of new bachelor's-level scientists and engineers. One of the report's findings indicated that the projected demand for scientists and engineers would far outstrip the supply, as older, Sputnik-era scientists retired and were replaced by fewer and fewer recent graduates in scientific fields. Even though the document was an internal draft and its forecasting methods were later criticized, its prediction of a looming shortage of scientists was made front-page news by then-NSF director Eric Bloch, who gave some 55 speeches on the subject between 1987 and 1990. The scientist shortage was amplified by several wide-circulation scientific journals, such as *Science, Nature,* and *New Scientist.*[25] The story was picked up by major newspapers, ranging from the *Christian Scientist Monitor* to *USA Today,*[26] and before long, college faculty and guidance counselors everywhere were spreading the news. Throughout the late 1980s, science and engineering careers were often presented to prospective students as a lucrative, secure, and even a patriotic future path. *Fortune* magazine, in an 1989 article entitled, "The Hot Demand for New Scientists," trumpeted the news of rising salaries for researchers, worrying only that there were not enough potential scientists to take part in the future bonanza.[27]

But a funny thing happened on the way to the future. The "scientist shortage" turned into a rather severe glut of PhDs in many science disciplines. The causes of the turnaround were several and virtually simultaneous. The end of the Cold War brought about a slow but inexorable shrinkage of federal defense laboratories and the research arms of corporate contractors building the Pentagon's new weapons systems. After a decade of nearly full employment, aeronautical engineers, nuclear physicists, and others working on the military buildup of the Reagan era found themselves expendable. The outbreak of peace not only constricted a large portion of the job market for new PhDs but also flooded the scientific job market with more senior, experienced personnel.

Meanwhile, corporate America began workforce layoffs and focused on short-term financial returns, beginning in earnest with the recession of 1990. Downsizing by firms in high-technology industries, such as computers, semiconductors, telecommunications, and chemicals, spelled doom for many research groups that had housed mid-career physicists, chemists, engineers, and technicians. In several rapidly changing industries, traditional research giants were radically restructured—the fates of the Bell Laboratory and of IBM's research labs are prime examples.

Before 1984, the American Telephone and Telegraph Company (AT&T) held a regulated monopoly on telecommunications in this country, and as such, was guaranteed a profit on its operations. A portion of these profits was devoted to supporting a large research institute, the Bell Laboratory, that was world renowned for its scientific discoveries. Much of the research at Bell Laboratory had little to do with the company's products and services, i.e., telecommunications; AT&T had the luxury of letting its research branch roam relatively freely in search of cutting edge science. In 1984, AT&T was stripped of its local telephone service network as the telecommunications industry was deregulated. Suddenly, the monolithic telephone company had to compete in a market, necessitating a hard look at corporate expenses, including research. Although spared for several years after the breakup of AT&T, Bell Laboratory inevitably was forced to shrink and to redirect its research along lines more in tune with corporate headquarters' immediate needs. Research in basic physics, for example, has been significantly scaled back in favor of product development.[28]

IBM has experienced a similar shift in its underlying business foundation, with comparable impact on its research and development arm. Like Bell Laboratory, IBM's Thomas J. Watson Research Center supported research groups working in many areas, from basic condensed matter physics to information theory. The center's achievements included two Nobel prizes. During the 1980s, however, IBM was unable (some say unwilling) to translate its success in large computers to the emerging personal computer markets. As a result, IBM entered the 1990s losing millions of dollars each year, and began radical restructuring to maintain its existence. One outcome has been a major contraction of IBM's research division, the first in its 50-year history. What is left of the research labs, predictably, is concentrating less on open-ended science and more on products to sell in the near term.[29]

Throughout the industrial ups and downs in the past, colleges and universities have offered a somewhat more stable, if often lower-paid, alternative to corporate labs. During the recession of 1990, however, the usually safe havens of academia offered decreasing shelter from the upheaval facing new scientists and engineers. Throughout the late 1980s, the population of traditional-age college freshmen was shrinking. According to the National Center for Education Statistics, by 1992, the number of high school graduates in the age range of 18 to 22 had shrunk to 87% of its 1982 peak.[30] Many colleges and universities struggled to maintain stable enrollments in the early 1990s, instituting marketing strategies to attract traditional and nontraditional students, including women and minorities.[31] Against this backdrop, science departments at colleges and universities were hard-pressed to avoid shrinking, much less grow at anything approaching their post-World War II expansion. Thus in spite of predictions of an upcoming faculty shortage, the number of faculty slots available to new science/engineering PhDs is not expected to grow for the remainder of the decade.[32]

In April 1992, a congressional subcommittee began hearings to investigate the scientist shortage that never happened. Testimony by National Science Foundation personnel revealed that the study forming the basis of the projections had relied on an extrapolation of scientist supply during the mid 1980s as a measure of future scientist demand.[33] In committee testimony, Peter House, author of the original study, explained that he had calculated the average number of bachelor degrees in science and engineering granted annually from 1984 through 1986—about 211,000 per year—and took this figure to be the minimum number of new scientists and engineers needed annually by the nation's economy. Working with demographic data, House then calculated that the supply of science/engineering graduates would shrink significantly due to the post–baby boom generation making its way through college. As it happened, House's base years—1985 and 1986—represented a peak in science degree production. By 1990, 176,000 science and engineering degrees were conferred, yielding a shortfall, by House's arithmetic, of 35,000 science and engineering personnel in that year alone. House's methods were met with criticism at the time; the NSF refused to approve the study for formal publication. However, after his predictions of scientist shortages were trumpeted by NSF director Bloch and repeated in the media, they rapidly became dogma.

As the employment picture for young scientists began to radically shift by about 1990, the notion of a scientist shortage proved difficult to dislodge. Well after the shortage began turning to glut, newspaper articles could be found extolling the virtues of science and engineering careers and bemoaning the lack of US graduate students in science programs.[34] The looming shortage of PhD scientists that never materialized became known as the Myth, and it spawned an insurgent group of recent and rather vocal scientists, the Young Scientists Network. YSN describes itself as "a loose-knit organization of young PhD scientists, science graduate students, and many other people who observe and participate in a discussion of issues involving the employment of scientists. We wish to inform government officials, the press, and the general public that despite widely publicized forecasts of an impending shortage of scientists, there is in reality an oversupply of young scientists." YSN has started a job-listing service and maintains a computer bulletin board with running commentary (and it does tend to run hot) on all aspects of the struggles new scientists face as they seek to advance in their profession.

Comments from young scientists, both on and off of YSN, suggest that the Myth and the Glut have had a most profound effect on their professional and personal lives. After being told that scientists and PhD engineers were in great demand, many new PhDs find themselves either in postdoc purgatory, where one respondent from YSN described young scientists and engineers as "highly educated migrant workers," or as placeholders in the unemployment line. The side effects of the Glut have been around for women for some time: in the 1970s, when universities were first made to comply with anti–sex discrimination laws, they hired women into newly created offtrack, temporary, and junior positions. A survey of 25 US universities showed that women faculty in science and engineering were *twice* as likely as men to be hired for dead-end positions.[35] Recent female graduates, relative newcomers to science, have overcome obstacles and endured chilly environments to make it to the doctorate level, only to find in many cases that there are few permanent jobs: It is rather like arriving at the celebration only to find that the cake has been devoured, and that she's expected by her hosts to be thrilled with the crumbs. This unexpected scientist glut has colored many women's perceptions of their chosen field.

Thea is disgusted at the double standard that she feels is applied to young scientists, as compared to the easier standards enjoyed by her predecessors: "I am getting so furious over senior scientists (1) criticizing young people for asking if they can get jobs before entering the field, and (2) accepting as OK the idea that people new to the field be paid poorly and work 12 hours a day, *and* move around the world every few years before being eased out of the field in their mid-30s after three or four postdocs. The latter suggestion was made both in a recent editorial in *Physics Today* and in the recent *AAS Newsletter*. These older white men spent 4 years in poverty to get their PhDs, then went directly to faculty. They got respect and position in the nonacademic community. They mistakenly try to apply their experience to what happens today, where [it takes] 6 years to get a doctorate plus 6 more years as a postdoc *plus* 7 years to get tenure. Hey guys, a woman is *too old* and exhausted to have babies then! And with the anti-intellectual bent of society lately, a scientist is perceived as one more person on the dole."

"When one of these tenured professors announces that he is resigning his tenured position to let some fresh blood in, or is going to go on contract (and yes, I know of one that has), or is spending his time training his graduate students how to teach even, then I will believe he is truly interested in the good of the profession."

The solution advanced by Marie to deal with the glut is to "Educate scientists broadly, so that they do not have only a single career path available (or visible to them) once they graduate. Excuse my bias, but scientists are some of the smartest people I know. They have extremely high skills in math, logic, science, computers, and reasoning. They are creative, yet can complete projects. Seems to me that a person with all those talents should be able to contribute to life in other areas—computer techniques, foreign language, medicine, teaching. Men and women of talent should be encouraged to study and do science, but at the same time, they should be told the facts of life quite early. I got into science fully aware that finding academic jobs would be very difficult, even for the top people. I was lied to by some people, but I knew they were lying at the time. A science undergraduate degree was never a waste of time for anybody—many people go on in other professions to be quite successful and fulfilled. A graduate degree in its current state is another story. One shouldn't go into grad school unless you like grad school itself. Otherwise, you might feel more than cheated at the other side. I enjoy what I'm doing now, this month. In 2 years I might be unemployed as an astronomer, but I will never regard this time I've spent as wasted."

The glut of highly skilled scientists and engineers resulting from defense industry downsizing is creating an opportunity of sorts for US schools. In late 1994, the US Department of Defense awarded $5 million to the National

Research Council to support a 5-year retraining program for displaced scientists who would consider careers in teaching at the middle or high school level. After a training period, an initial class of 20 "teaching fellows" will begin teaching full-time in inner-city Los Angeles secondary schools. According to a National Research Council press release,[36] the anticipated outcome of the retraining process is that these new teachers—who will bring to the classroom years of real-world experience in applied science and technology—will positively influence the large number of minority students in their Los Angeles classrooms to pursue careers in science, mathematics, and engineering. Exposure to experienced scientists is great for kids, whose facility with math and science may be bolstered. Proficiency in these subjects boosts scores on college entrance exams and may open doors to colleges and academic scholarships. But does anyone see a problem with the explicit goal of encouraging students to pursue scientific careers, a motive with the implicit promise that good jobs exist? The situation is comical: Twenty out-of-work scientists and engineers, being paid with tax dollars to encourage more people to become scientists! Promoting a more science-literate student population is of course a wonderful motive, but people of color and women must be accorded the truth about future careers in science. Perhaps I'm being too conservative; perhaps in the year 2007, when the fresh PhDs from the high school class of 1997 are about ready to job hunt, the entire picture will have changed. Today, I'm not sure I'd bet even $5 million on it.

Leah, a professor of mathematics, responding from the Internet, is a rare winner in the glut games: at 27 she's landed a tenure-track faculty position at a top university. Still, she's aware of the problem and sympathetic to her peers: "Ugh. I find it really depressing. I feel obliged to be realistic about the job opportunities to people thinking of entering the field. I hope the good ones aren't discouraged . . . . I don't know what to do about it. I would say, however, that this is a major problem for generation X in all fields: science used to be exempt, and it's still a little bit better than some other professions. You can't tell me you are more employable as a PhD in classics, or as a painter or an opera singer . . . . In general, we need to find jobs for the 20-somethings. I am one of the few lucky ones."

Carol, the computer engineer, sees the direct and open approach as the best way to ensure that the best people continue to enter the sciences, without painting an unrealistic picture of employment prospects. She also places part of the blame for the glut on scientist's own shoulders: "Anyone who deeply loves science and who has the talent to do good science should be strongly encouraged to study science, regardless of the job market. To do otherwise would be criminal suppression of the person's natural gifts and inclinations, a denial of who they are, a devaluing of their humanity. At the same time, unrealistic promises about rosy futures should not be made. The realities of the various job markets should be made clear, so that the person has a firm and true basis for making decisions . . . . I think the only long-term solution [to the glut] is to better sell the everyday benefits of both pure and applied scientific research to the population at large. Except for pork barrel projects, legislators tend to not support science funding, because their constituents do not support it . . . . Within corporations, it is also necessary to improve the sell job to those who control R&D funding. The bottom line is that each individual scientist should be able, at a moment's notice, to justify their work in terms of its benefit to society and/or corporate profits. If we don't take responsibility for this, who will?"

Vicky redefines the problem: "The problem is not a glut of scientists—if the United States is so full of them, then why are we behind in technology? The problem is lack of support from government funding and society. In my field [astronomy], I believe that the 'glut of scientists' will ease as the economy improves. In good economic times, people feel the need to explore and develop science. In bad times, society begins to question why money should be spent on the 'unnecessary luxury' of science. Educating politicians about the need for science to solve economic and environmental problems is part of the solution."

Margaret, who writes from her perspective as a software executive, says, "All through my years working on my PhD I was told that for a computer scientist, it was a total waste of time; that I should stop at a masters. And I was told I would be 'overeducated.' One woman went so far as to say that women tended to be 'overeducated' because we were afraid to enter the job market! But I stayed in, because I was/am in love with the field and the beautiful order that science brings to it. Upon graduation I learned two things: (1) A PhD is exactly what you make of it. I do not see a 'glut' of educated people—I see a 'glut' of college teachers—I have never tried to get a job teaching (lots of reasons for that). (2) Getting my PhD was the best thing I have ever done for myself—it has helped me enormously with my own sense of my credibility, because I really am an expert in an important area."

Viewing the job scene from her standpoint as a computer scientist in an industrial lab, 49-year-old Harvard-trained Nora now debates her practice of encouraging women to enter the sciences. "[The glut] disturbs me. I have worked in the past on encouraging women to go into science, but it's not clear that this is a good idea anymore . . . . [however] I would urge other women not to be discouraged by the attitudes of people around them. If you really want to pursue scientific research, just do it."

Women and girls with scientific aptitude are at a particularly difficult juncture in their advancement toward full professional participation. What is the correct path to take in order to ensure that their training is useful and that their career expectations are met? At what point is the emphasis on recruiting women into the sciences intended more for the good of the institution, which may meet its affirmative action goals and benefit from bloated overhead payments, than for the good of the student? Can we guarantee that science jobs will be available for women scientists when they graduate? As some of my interview subjects noted, they feel far better off having science education and training under their belts than they would if they had avoided the field because of lousy job prospects. They feel that the knowledge and skills they have acquired through the scientific apprenticeship can be reapplied to other fields with excellent outcomes. But is this a fair deal to offer women who have been in college for 8, 9, or 10 years or more, learning to be scientists, not administrators, high school teachers, or stock analysts—fields for which a sound 4-year liberal arts education can produce outstanding performers?

At question is a larger public policy issue: Should taxpayers, through federal grants to university research labs, continue to subsidize the training of scientists who will never apply their expensive training directly to the sciences? Or, should those who choose to enter the field be expected to support themselves throughout their educations, as nonscience students are expected to do? An honest discussion would be welcome, both in Congress and by national science policy organizations, about how science benefits the United States and how those benefits should be paid for. If a consensus can be reached that basic scientific inquiry merits national funding, then funding should be increased to meet the level of qualified personnel that were trained with federal dollars. If it is agreed that scientists and engineers contribute significantly to technological innovations and economic vitality, then, to the extent that government plays a role in private business, it can influence the level of spending on research and development. The scientific glut presents an opportunity to analyze the ramifications of the collapse of the federal defense industry, with an eye to how we might retool the national science machine. Inherent in this discussion are both opportunities and pitfalls for prospective scientists of both sexes, to whom honest analyses and forcasting are owed.

3

Differences in the Lab
Women's Scientific Styles

Do gender differences make a difference in the workplace? Sociologists have pored over the implications of sex differences in the work environment since women entered it. They have looked at how women and men work together, how women respond to working in a male-dominant situation, and how the influx of women into previously all-male fields has affected the attitudes of the men they're joining. Research includes the study of tokenism, wherein a sole or small number of representatives of one group are incorporated into the fabric of a dominant, mostly homogeneous group—a scenario played out daily as women enter the sciences.

Rosabeth Moss Kanter,[1] now a professor at the Harvard School of Business, looked at women's behavior and treatment when involved as tokens. Kanter showed that token women are subject to distinctly negative actions by the dominant group, including performance pressures, whereby the minority member is held to much stricter standards of performance than are majority members; polarization, wherein the dominant group emphasizes its culture and its dissimilarity to that of the token; and stereotyping, where token women take on, or are saddled with, a number of exaggerated roles. She gave these stereotypes colorful but descriptive names: mother, seductress, pet, and iron maiden, each referring to the behaviors that women acted out within the group of men surrounding them in the workplace. A token woman "mother," for example, might sew on buttons for her colleagues; a "seductress" might form a close alliance with the dominant male in the group, perhaps in order to gain his "protection" from the other males; and a competent woman, who expected equal treatment within the group, might fall prey to being cast as a tough or even dangerous "iron maiden," not to be messed with in any way. These stereotypes probably seem familiar to most people who have worked in mixed-gender settings, even if they are untrained in the sociology of workplace dynamics. From Kanter's research it seems that both the female token and her male co-workers have a part in perpetuating the roles, which provide neat boundaries and expectations for the behaviors of both the women and the dominant male group; in other words, a survival mechanism in an extremely stressful situation. Kanter shows how all of the phenomena resulting from extreme imbalance in gender makeup of working groups create handicaps for the token workers.

Kanter's theory of the negative effects of performance pressure has been applied by sociologists Stacy Rogers and Elizabeth Meneghan in a study of women's lack of persistence in pursuing male-dominated undergraduate majors, including science and technology.[2] While their research results did not completely support Kanter's thesis, the authors reported in the December 1991 issue of *Gender and Society* that performance pressure had a "significant, negative effect on likelihood of persistence." In other words, where female students were enrolled in majors where they were the distinct minority, the pressures they experienced from male colleagues were likely to contribute to their dropping out of the major.

Amy Wharton and James Baron studied the effects of gender mixing on employee well-being, and noted, in the *American Sociological Review*, a drop in job-related satisfaction and self-esteem by both men and women where numbers of employees of each sex become balanced, rather than male- or female-dominated.[3] This is interesting research, running counter to the idea that gender relations would mature as relative representation of each sex became equal. It is also useful to compare this research with Kanter's, where it would seem that dominant/token groups have developed survival mechanisms that evenly balanced groups cannot. When gender groups reach numerical parity and approach equal strength, the more level playing field would appear to lead to a protracted workplace war between the sexes. This could have unforseen outcomes in science when women are as well represented as men.

The sociological study of gender differences and group dynamics can help to explain the incidence of gender conflict in recently mixed fields and of disparities in participation by sex in the scientific sphere. Another area that should be addressed is whether the frictions existing in today's scientific workplace, which may lead to women's and girl's attrition from the sciences, are due to some fundamentally different styles employed by its practitioners. Do behavioral "styles" affect scientists and their work? To accept the question's premise, one must first agree to the concept that people "do" science differently; meaning not necessarily a difference in research methods (though this may be a consideration), but rather that variations in habits of personality and professional style exist and that they affect one's work and colleagues. To posit the question further—to ask if women and men might "do" science differently with some gender-specific similarities—one must examine the possibility of shared gender traits among scientific participants from each sex.

Here is where one begins to see the culture of the scientific endeavor in a rather unbecoming light. As I discussed the concepts of work styles and scientific success with my female and male respondents, the hue and cry coming from younger scientists was almost deafening: even though they may themselves enjoy a relatively mature group, or supervisor, or institution, young scientists are appalled at the overarching acceptance of childish behaviors that are present at many levels of established science. Science, as practiced in the United States and much of the West, is a game in which the winner's circle is constricting. The behaviors that develop at the gaming table reveal much about the players (scientists) and the manufacturer of the game itself (science funders). My respondents are emphatic that a change must occur, and it is young women whose voices are the strongest.

Do Women and Men "Do" Science Differently?

I put this question to subscribers of a number of electronic scientific discussion groups. If indeed women are to change behaviors that they see as inappropriate in scientific culture, it would be helpful to know how the behaviors broke down along gender lines. The responses to this question were quite varied, as we'll see.

The potentially positive outcomes of applied feminist standpoint theory were described by my first respondent: "I believe that women have a separate set of preconceived notions than men do about science," says Sharon, a computer scientist with NASA. "Men do the brute force approach to science, where women tend to be more cautious. I believe that having teams made up of men and women is the best way to do science because you will get these different perspectives working together. Make it even more diverse . . . add minorities."

Marie disagrees. As a 31-year-old postdoc in astronomy, she says, "Everyone has different styles, some may be considered more 'feminine' than others, but in my experience, I cannot identify a 'female' way of doing astronomy. There are more male jerks, but there are more males, period. If there is a difference, it might be in presentation style, rather than the actual 'doing' of science. Men might be more aggressive and assertive in the presentation of their work—everyone knows the 'macho' man approach. But even more men don't really respect that way of speaking. Assertions over careful arguments may win in business, but it gains few fans in my world. We often laugh at them behind their backs. This again may appear more often in men because there are a lot more men, and therefore the variety of observed characteristics may be larger."

Carol, a 41-year-old computer engineer with an interest in cybernetics, places the difference on "the perspectives that are brought to bear on the process. Men tend to view science as a contest on par with all other one-upmanship games. It makes little difference to a man whether he 'wins' a science game, e.g., by having his name listed first on the publication, or whether he 'wins' a car game, e.g., by owning the sportiest model, or whatever. For a man, winning the game is all that's important. The nature of the game may influence the level of increased status that can be gained by winning, but otherwise, it's considered unimportant. Women, on the other hand, do not view science as a contest, much less as a means for acquiring status at the expense of others. Women value the scientific process and its results for their own sakes."

It's easy to see surface differences between the sexes, since mainstream American culture, despite its claims of gender egalitarianism, defines very specific behavioral roles for men and women. Fashion trends are an inconsistent and superficial but nonetheless striking example, particularly since no one else controls our personal clothing decisions. With that in mind, whither the comfortable, unisex clothes of the revolutionary 1970s era? Modern 1990s women are flocking to the stores to purchase girdles and padded push-up bras,[4] and little girls are dressed in frilly, pastel togs (today's parents even snap beflowered and beribboned pink headbands around their infant daughters' bald pates). Menswear has remained essentially static: comfortable shoes, comfortable slacks, comfortable underwear; the ensemble routinely made uncomfortable only by the addition of a necktie. Boys' clothing is covered with dinosaurs, sports images, and the color blue. Barbie dolls are de rigueur for girls, while boy's action figures (never dolls) are forthrightly murderous. If consumer and fashion habits have anything to do with a celebration of sex-specific behavior, then we have to accept either Americans' desire for gender difference or its reality. After all, though we may be victim to a superb marketing job, *somebody* is buying these things.

Besides the evidence one can see in closets and on the playground, we've noted a couple of feminist theories that allow the possibility that women may share behavioral commonalities, as defined by difference feminism; and that women or members of minority groups develop experience-driven outlooks, as described by feminist standpoint theory. Linguistic assignation of gender traits by the dominant culture, as value-laden as they may be, are revealed through adages such as, "women's intuition," "it's a man's world," and "that's just like a woman."

Women and men are even defining themselves as different, by developing and attending professional seminars and buying bestsellers on communication differences, by spending their vacations with same-sex groups on wilderness adventures, or by looking for their "wild man within" during men's drumming weekends. Strange as it may seem, women who came of age during the cultural upheaval of the 1960s and 1970s are now toting their daughters to charm school. An analogous class for boys, to learn poise, grace, and delicacy in servitude isn't offered. Though certainly never true for every individual, it's fair to say that through some combination of genetic heritage and cultural indoctrination, women and men in Western culture have developed, or tend to exhibit, some degree of gender-specific behaviors.

Schools do much to reinforce and encourage specific behaviors. According to *How Schools Shortchange Girls*, a report issued in 1992 by the American Association of University Women[5]:

> Across the whole spectrum of the K–12 curriculum there is currently more emphasis on the development of assertive than affiliative skills, more reward for solo behavior than collaborative behavior, more reward for speaking than

for listening. The curriculum can be strengthened by consciously focusing on the development of reflective, caring, collaborative skills as well as those skills emphasizing individual performance and achievement.

It's accepted that successful scientific work requires basic qualities, among them intellect, memory, reasoning and organizational skills, and persistence. Also valued by scientists, but not necessarily required, are curiosity, creativity, and a love of one's work. Going beyond these basic traits, there is large variation in scientific personalities, just as there is in the general public: some patient, others insistent; some gregarious, others reserved. None of these characteristics seems to reside more predominantly in either sex.

What characteristics differentiate male and female scientists? Indeed, do any exist at all? I asked this question in several different ways to subscribers of a number of electronic discussion groups, as well as to women I interviewed personally, and received enough responses to convince me that blanket generalizations about sex-based behaviors in science are ridiculous. It is not ridiculous, however, to point out that some behavioral characteristics seem to have become part and parcel of scientific culture, and that among these behaviors are several considered risky, counterproductive, or juvenile when exhibited in social settings. What galvanizes the opinions of female and male respondents is the notion that men are often rewarded for using these behaviors in sporting or business contexts, and that their transference to the sciences makes much more sense to men than to women, for whom the behaviors must often be learned. The specific behaviors we'll discuss are competition and intimidation, which were among those brought up most frequently by respondents as having a definite presence in scientific culture. Next we'll look at a stereotypically "female" trait—intuition—and see if and how it is applied to science. To complete the chapter, we'll look briefly at the ethical or moral decisions facing scientists, and whether men and women approach them differently.

Competition versus Cooperation

My respondents often mentioned science's competitive working culture as having significant potential for negative outcomes. Competition, as defined by Webster's, denotes "a striving for the same object, position, prize, etc., usually in accordance with certain fixed rules."[6] Competition is often assumed to involve a scarcity of resources, or prizes, as in competition for survival among animals dependent on a limited food supply. In today's science experience, with jobs and funding in short supply, it would be fair to assume that conditions are ripe for competition.

In science, competition occurs in a variety of instances, starting in early educational preparation: testing, grammar school contests, and high school honors prizes, and continuing through graduate school, with applications for entry into prestigious groups and continued academic competition for grades affecting each student. After the newly minted PhD enters the professional world, competition becomes more intense. Job applicants must rise above hundreds of others for solid academic positions. Applicants outside of the academic sphere must compete against other PhDs as well as lesser-trained workers, who, by the limited but adequate nature of their preparation, bring down the going salary and increase the numbers of competitors for available positions. Competition among increasing numbers of scientists for limited quantities of research funding is also intense. Some US scientists have the recently added pressure of competing in a global market on the home front: Survey results from 1991 showed that international scientists made up more than half of math and engineering graduate school students in US institutions.[7] In addition to students, fully trained international scientists are immigrating to the United States with permanent visas from Asian and Eastern-bloc countries at an unprecedented pace. Labor certification data for 1991 and 1992 show India as the country of origin for 15.7% of all science and engineering immigrants, with China making up 13.5% of admissions. Between 1968 and 1990, the number of immigrant scientists and engineers admitted to the United States on permanent visas hovered between 6,400 and 13,300 annually. In 1992 alone, the number admitted was 22,900.[8]

Research scientists are pitted against one another by personal drive and the expectations of funders. The thrill of scientific discovery leads many researchers to do their utmost to be first out with a paper; for others, it's the cold fact that employers require papers and that funders expect results in order to continue funding grants that leads them to churn out data. In science, there seems to be no shortage of areas of study or discovery; rather, the scarcity that is implied by scientific competition seems most directly related to ego and money. Hot scientific groups get lots of attention and lots of grants. Lesser ones do what they can to keep their foot in the door. Unproductive groups don't often last; and those on the fringe, working perhaps on totally new ideas, had better come up with some popular insights fast or they're likely to toil in underfunded obscurity for a long, long time.

The behavior that counters competition is cooperation, and it is a trait that women have been attributed with for ages. The concept of cooperation plays differently with different audiences. Some consider it nothing more than a cultural nicety, born of easier times that allow such things, and others think it is a vital trait in human survival itself. Recent scientific theories champion cooperation as a more beneficial behavior in species survival than competition, providing a completely different perspective on Darwin's "survival of the fittest" notion.[9]

When Linda Wilson, president of Radcliffe College and herself a chemist, told an influential group in 1992 that the new generation entering science, particularly women and minorities, would not stand for the current culture of "ruthless competition and single-minded devotion to the job at the expense of everything else," her audience couldn't help but notice. Some in the audience, including MIT astrophysicist Bernard Burke, interpreted Wilson's remarks as intended to remove the competitive spirit from science, according to an article by Jeffrey Merves[10] in *The Scientist.* Burke objected to the idea, reciting the well-worn battle cry: "You don't get to do the best science by being a nice guy. Opportunism and competitiveness are essential" to the scientific enterprise. Let's listen to what survey respondents had to say about scientific competition versus scientific cooperation.

A female postdoc who responded to my query on the Internet said: "We are all competitive at some level, particularly as the job market is so bad. It tends to be 'anonymous' competition, in that I have never been in the situation of feeling like someone I knew was competing with me, but I feel competitive with the others out there that are applying for the same positions. I think [women scientists are more cooperative than men]. I am, for the first time, in a laboratory situation, having always been a field researcher prior to this. The laboratory I am in is run by a woman, and most of the students are women. I am the only postdoc. The principal investigator has had, and continues to have, male students, but they are often "cooperative" men. In general, the sense is of great support, and if someone cannot do something, or needs help, there is usually a way to work things out."

From Karen, an assistant professor of mathematics at a West Coast technical college: "My limited experience suggests that *really* successful scientists (mostly mathematicians, since that is my field) who are also reasonably sane (a significant caveat) are not particularly competitive or intimidating, regardless of gender. Among my peers, men do seem more likely to be competitive—though not all of them by any means."

Regarding whether women scientists are more cooperative than men, this same respondent provided a personal example: "Six members of my department were supposed to meet. Two showed up, one did not get a message changing the time and showed up later, and the other three did not show (I was one of this group). It was registration week; there were students in and out of my office the whole time; and I simply forgot. The two who made it to the meeting have the offices on either side of mine. Neither one knocked on my door on his way to the meeting. Our department is very compact—it would have been the work of less than 60 seconds to round us all up. Instead, the person who had called the meeting sent an e-mail memo to the department head (copied to the rest of us) detailing the absences. I cannot count the number of times I have knocked on that particular faculty member's door to remind him of meetings."

Another scientist, a 34-year-old engineer working with the US Environmental Protection Agency, polled her colleagues about the competition question. "I asked many of my women co-workers whether they thought there was a difference between men and women scientists. Most of them said yes—and during discussions in which we tried to explain the differences, the words 'cooperation' and 'competition' kept coming up. In general, I think women strive for a cooperative workplace, whereas men seem to fuel and feed off of competition. One of my co-workers, a woman geophysicist, told the following story. She was leading a five-person kayak trip. For lunch they stopped at an island. Two of the women were playing Frisbee. The man stood equidistant from each of them and threw rocks at the Frisbee! For her, this symbolized her academic and working relationship with men—the women would try to cooperate and 'play' together nicely, while the men would disrupt the game. I agree.

"This difference in approach has always been apparent to me during technical meetings. The men talk too long, blow their own horns, isolate themselves from the group. The women, in general, work toward consensus. For *years*, I worked in predominantly male environments and I felt extremely out of place. I would never talk during meetings. I didn't think I had what 'it takes' to succeed in consulting. Now I realize that my style is just different. And when I have a chance to run my own meetings, I run them *very* differently than my old bosses did. And I *am* succeeding as an engineer, but I have had to find a different environment for myself."

Cindy, a grad student in physiology, read meaning into my question about scientific competition: "What makes you think that women dislike competitiveness?" she responded. "Competition (to do the best science) and cooperation (to work together to accomplish more than we can individually) are not mutually exclusive. I am a woman, I believe in competition, and I enjoy cooperation with other scientists. May the best idea win!"

Most of the women I spoke with stood somewhere along a continuum, from not being competitive by nature, to having learned to be competitive to get by in the field, to loving competition and gleefully jumping into the fray, confident of emerging victorious or at least knowing and accepting the maxim that somebody has to lose. Only a few women believed that competition was an entirely negative behavior. The question of the place of competition in our lives and work has yet to be answered. Feminist philosopher Helen Longino [11] addresses the difficulties faced by women who want to be mutually supportive in our efforts toward full participation in life, but find it difficult in a world where competition is the name of the game:

> We are often faced with a conflict between our commitment to, indeed our longing for, solidarity with other women and our need to compete in the marketplace for work. Without a better understanding of the competitive structures in which we work and play and of our own responses to these structures, we believe most women will remain frustrated, guilty, angry, and divided. . . . If we could stop feeling defensive and fearful long enough to consider how we compete not only for money but also for attention and affection and righteousness, we might be better able to eliminate the negative elements of competitiveness from our lives.

Longino encourages women to step back and consider how competition controls their lives in order to fully understand how it leads to negative outcomes. Women and men, at least in Western culture, have been indoctrinated to see competition in a favorable light, particularly through sports. Many girls are now encouraged to compete and play hard in sports, and for the most part, boys are simply expected to compete. This situation alone may account for much of the approval toward competition that my respondents displayed. The concept that competition always involves a loser does not equate to a bad outcome; we like to think that losers somehow deserve their comeuppance: either they weren't prepared, performed poorly, or simply don't cut the mustard; or on the other hand, perhaps we find that the winner was so superb that we could rejoice in seeing their skill. In school, in sports, in social relationships, in business—competition rules, and American society sanctions it.

The responses I received from male scientists on the subject of cooperation versus competition are fascinating, so I've included quite a few of them. I asked over the Internet whether men reward competition in science, and whether competition is in fact the accepted modus operandi of scientific work and culture.

Jack, employed full-time as a senior research fellow in physics, says, "Part true and part false. Effective competitors are rewarded, because if there are indivisible resources, and several people or groups want them, those who compete get what they want. I do not believe that the resources should be as indivisible as they often are made by grant-providing and -receiving institutions, but what can I do? Most scientists do not particularly like the explicit competition of grant seeking. They want all qualified research to be fully funded. Within research groups, cooperation is far more the rule than competition. Any competition for pieces of research to work on are easily settled by seniority, but it has been my experience that people work on what they want to, within the contractual framework of proposal and grant. Competition to produce interesting results is friendly. There is plenty of room for those who dislike competition."

Many of my respondents pointed out the scarcity of funding for the good research that is being proposed. Because of ever-growing numbers of scientists chasing after static or shrinking amounts of research funding, scientists who might otherwise eschew competition may find themselves in a daily contest with their colleagues on several levels, even though this was not Jack's experience. Competition can occur within research groups, where grad students and postdocs compete for greater responsibility or equipment access; sometimes it takes place among various institutions and the labs housed therein; and quite often, research groups compete internationally. One of my respondents, an astrophysicist, pointed out that competition can have "dangerous" outcomes, since funding and other decisions are often made by relatively few people, and the field can end up "being dominated by the egos" of a very few scientists who are placed in positions of significant power.

Charles, a biologist working as a postdoc at a prestigious research university, made it clear that not all men find competition to be a positive force. He pointed out that in his experience, cooperation and collaboration have been much more constructive. In fact, much of the competition that Charles has seen ended unfavorably—producing "sloppy work." Certainly there is an element of haste implicit in competition: getting there first is part of competitive culture. In science, where scrupulous attention to detail and accurate work are essential, speed may not equal success; yet whoever gets results first is clearly rewarded by the system.

The idea that competition may diminish the quality of results was echoed by Jerry, a chemist: "Competition in science is really very minor, with the glaring exception of funding. However, the ability to produce high-quality research and proposals is quite an individual thing. Politics are quite vigorous—bordering on the insane at times—but talented research is talented research, and nobody can out-duel you at that. The key here is the old phrase, 'work smarter, not harder.' If you try to beat somebody out with the sweat of your brow, you will most likely lose to somebody who has thought things out a little better, and that person is likely to be working on something you haven't thought of anyway, so if you beat them once, you probably missed the boat, and have gained little if anything."

A male assistant professor of chemistry believes that men reward competition, but doubts that it ensures the best science. In his opinion, competition "taken to extremes" encourages fraud. He continues, "I have known of labs where postdocs were put in direct competition with each other. This led to ill-feeling and a lack of cooperation and I'm sure did not improve the quality of science being done. On the other hand, I think some level of competitiveness is an essential driving force for most people. I do think that if we were training students correctly, most of that competition would be with themselves."

Michael, a high-energy physicist, alluded with faint praise to the high-powered personalities found in the field: "I wouldn't exactly say that men 'reward' competition in the sciences; it's more that most people in the sciences seem to just expect everyone in the field to have competitive, aggressive tendencies and be workaholics. I personally am not like that, and am hoping to survive in the field by finding like-minded people to work for and with, and to earn the grudging respect of the aggressive types by proving to them occasionally that ignoring the pressures—giving oneself time to think things through before acting—can pay off."

Finally, Dimitri is a true believer in scientific competition. A solid-state physicist and a new father, he believes that competition is the only way to ensure that the best science is being done. In fact, he thinks that the academic structure needs to be reformed to allow the best and brightest young scientists to compete with established faculty for influential positions. Dimitri takes the unique point of view that the tenure system "heavily discriminates against much more qualified younger applicants . . . . Since many older scientists were hired during more favorable times, most of these older scientists wouldn't stand a chance in today's job market." In Dimitri's science, the "managed economy" of academic job protections would be replaced by a free market, tenure would be eliminated, and science professors would face a lifetime of jobs competition. Not surprisingly, when I asked him about his goals and interests, he responded, "Getting out of physics," and "wealth," two goals which many of my respondents seemed to find connected.

The men and women I corresponded with agreed to several points regarding the competition question. Both were likely to concede that at least some competitive element is a positive feature in scientific inquiry; that it jump-started the quest for knowledge and made individual scientists and research groups function at peak performance; and that it provided a mechanism for weeding out poor proposals and shoddy research. Women and men also agreed that competition could have negative outcomes for scientists and for the field, most markedly when competition forced important research proposals out of the game; when scientists spent more time preparing proposals for intellectual showdown than performing research; and when the intensity of the competition lead to corruption or fraud. For most scientists, competition in their own work was simply expected; but those I spoke with indicated that competition was becoming a more negative force than positive.

Though I don't believe that they are alone in their view, the women I encountered were more vocal than men about the value of encouraging a cooperative, supportive model for scientific inquiry. I find this surprising, because the women that have succeeded in becoming scientists have had to be particularly skilled competitors throughout the process. From my conversations, I'll predict that women will have significant impact on redesigning the characteristics of science that make competition inevitable, putting in their place rewards for cross-pollination of ideas and cooperative work.

Intimidation

In talking with scientists about competition, they often mentioned intimidation as another behavior for which science is notable. I began asking, again in personal conversations and via the Internet, if intimidation was an acceptable behavior in science, how it was used and by whom, and whether or not the respondent used intimidation. The picture painted by respondents shows a field where intimidating behavior is common, not only on the part of the powerful to the less powerful, but even among equals as a part of competition. Intimidation in science can show up in the peer review process, where anonymous reviewers can tear a paper to shreds; at scientific meetings, where some respondents described the aftermath of a presentation as similar to being thrown to the wolves; and in the classroom, where as we'll read later, some professors seem to enjoy terrifying their freshman classes with predictions of mass failure. Unlike the occasional positive response I heard to competition, very few women I corresponded with felt that intimidating behavior was beneficial to science. Though they found it repugnant, those women who used intimidation determined that it was a behavior that was understood by their colleagues, and that by using it they were more effective in the workplace.

Gwen is a 23-year-old second-year grad student in space physics. She states that, "Intimidation is never an acceptable behavior in any facet of society, including science. This holds true in one-on-one relations between a mentor and advisee, in the classroom, in the lab or observatory, and in public at meetings.

"I think that most intimidation comes from the way science is taught and perceived by all those involved. Scientists always think that they are the brightest and best of society, so they have to 'test' and make newcomers 'pay their dues' and go through a lot of unnecessary stuff just so that thay 'know what it is to be a scientist.' This is also used to weed out the low and average students . . . . in an introductory physics class at a university that has a very strong and difficult physics program, the professor comes into class the first day and the first thing he says to the class is that by mid-semester half of them won't be there and that maybe a third of the class will make it all the way through. This sets up a hostile, aggressively competitive atmosphere. Of course, I have also had a female faculty member tell me to rethink what my career plans were because I didn't do well on my first physics exam my freshman year . . . so female scientists also do this."

In her discussion of intimidation, Gwen noted her perception that the behavior is being used against women by men in the hope that women will find the field so inhospitable that they'll drop out. This is certainly true in her case. She had been told that if she wanted to be a "real" astronomer she would be forced to forego a normal social life, would not be able to get married or have children, and would face additional inevitable constraints on her life as a scientist. Even the person she identifies as her mentor had bad news for Gwen, telling her that "not wanting to work late at night by myself in an observatory in a remote part of campus for safety reasons was not a legitimate concern and that I should just 'buckle up and deal with it.'"

Gwen pointed out the common episodes of professional jousting that occur at professional scientific meetings, where she noted that, in addition to the normal collegial questioning that occurs in response to scientists' presentations, some people go overboard trying "to find fault and errors in the work being presented," particularly if the presenter is a young scientist. In a blatant example of both intimidation and sexism, Gwen recounted when she was told at a meeting that it was not her work that attracted people to her presentation, and that she "should be happy that *something* does."

Science does not come off as a particularly welcoming place for women when seen through Gwen's experience, and Gwen's stories were not at all unusual among the women I spoke with. Somewhere during its development, science took on a decidedly combative overtone. Why? Are some scientists overwhelmingly insecure—a vestige of being intellectually overperforming outcasts in their youth? Does the influx of women into the field make male scientists more concerned with laying claim to what's "theirs," rendering the spreading of fear among neophytes a useful activity? Is intimidation science's version of professional hazing? Or is it just a blustery means for maintaining high enough standards for those allowed into the field? Whatever motivates scientists to intimidate does so at the expense of turning off potential science workers, who may become so upset by the pressure that they avoid the field altogether. Certainly educators and administrators should monitor the climate in their insitutions to ensure that students do *not* get the kind of messages that Gwen received.

What may pass for intimidation in some places is seen as a healthy part of scientific give-and-take in others. Much of one's view depends on one's own standpoint; and women, apparently, are not as familiar with a culture full of boisterous ribbing as are men. Vicky, a second-year grad student in astronomy, agrees that intimidation is unacceptable in science, but that a milder form of banter is okay. While Vicky does not use intimidation as a tool, she did point out that she is trying to become adept at the playful conversations, full of good-spirited (but often quite pointed) teasing, that she says take place regularly among physicists and astronomers.

Intimidation is distinguished from debate by Carol, the computer engineer whose views we heard earlier: "Intimidation is never an acceptable behavior. It is also self-defeating in the long run, since the person(s) being intimidated will likely seek revenge, if and when the opportunity presents itself. It is important, however, to distinguish between intimidation and adversarial debate over scientific methods, data, etc. The former is one-sided and never leads to positive results. However, the latter, while sometimes causing painful humiliation, is a necessary mechanism for establishing the level of truth and reliability in scientific work. The negative aspect of such debate is the male tendency to transform it into a win-or-die competition."

Forty-one-year-old Thea, a grad student in astronomy at a national observatory, thinks intimidation is "absolutely" accepted in science, and that it works very well in its intended oucomes. She further points out that intimidation, when used by women, is particularly effective, since people are unaccustomed to seeing women use it. "I've seen powerful scientists use [intimidation] to deflect criticism and to determine if someone is bluffing them. I don't use it in my work yet, but I suppose I will. When women use it, it just terrifies men."

I heard from another woman who admitted that her own behavior is sometimes seen as intimidating, primarily because of her comfort with argumentation and her assertive nature. While she did not intend to come across as intimidating and while she pointed out that she was in no professional position to intimidate anyone at her institution, she still felt that others responded negatively to her. Perhaps assertiveness and confidence are still not acceptable behaviors for women, or perhaps we are simply not accustomed to finding these traits in women who are not otherwise objectionable. Many women alienate those around them with their single-minded power plays, no differently than men who use these tactics. Are we conditioned by these experiences to react negatively to any woman who speaks her mind? Think about the public outcry when First Lady Hillary Rodham Clinton began discussing her opinions on national issues. She spent the next 3 years of her husband's presidency apologizing for her outspokenness and backpedaling from her obvious command of the issues. The public just couldn't deal with an obviously intelligent, articulate, and opinionated woman so close to the power.

Carla, who offered that she does research in the most "macho" part of medical school, the department of surgery, says that intimidation is "quite a real issue. I get shot down in review processes because my concepts are not yet well known, for mostly silly reasons. But I don't count failures. No one does. Intimidation is more like coercion to think like the group; group-think. Outsiders [are] not welcome."

Perhaps the level of intimidation used in science varies in intensity from field to field, as suggested by Carla's reference above to her ultramacho department, and by a woman scientist who doesn't find intimidation to be a problem at all. Jeanine, doing a postdoc in comparative physiology, says that, "In groups, like at meetings for instance, I am often the only woman in my group, and I can't recall ever feeling that someone was trying to intimidate me. . . . I have found the culture of science to be supportive, encouraging, and helpful. That is how I would describe the *vast* majority of scientists I have met, and it is just as true of male scientists as female scientists. There are a few arrogant idiots in science just as everywhere else—and my experience is that arrogant people are no better liked in scientific circles than they are in society in general. My experience also is that some of the most famous science "stars" are some of the nicest people."

Again, let's sample a selection of men's responses. Once more I am providing equal time to both sexes, because the reactions of men to the attributes of a field that their sex controls are of particular interest. After all, if men set the rules, then men accept the rules, right? Not always. Juan has the perspective of being from a Latin American country, where he was trained in geophysics. In the United States on scholarship to complete his PhD, Juan says that intimidation "is the name of the game in academia. In fact, it's a formal academic quality. In my view, and many others', the academe can be accurately defined as a continuous confrontation of egos, as most academics are egocentric, intellectual bullies. We students have to face, undergo, and witness a lot of these confrontations and cope with the intimidation they can produce. Faculty members only point out what's wrong with us, but not what is right."

Juan's views are not isolated because of his international perspectives. US-born and -educated men are in full agreement. For example, some areas of astronomy have become riddled with intimidating behavior, says Skip, an astrophysicist at a prestigious research university. Skip thinks that these areas have come to be dominated "by the ideas of a small group (or even single individuals), who are quite intolerant of any work that can in any way be construed as implying that something they've done is either incorrect or incomplete . . . . As you can probably guess, I find such situations completely unacceptable. Their only effect is to squelch creativity."

Jack acknowleges that intimidation occurs within science, but doesn't think that it is used any more often in science than in other employment situations. He concedes that communication styles can often lead to an assumption of intimidation on his part: "For some reason that I cannot understand, people often find me to be intimidating when I have no desire to be. They find my tendency to state opinions without lots of disclaiming words like 'I feel that . . .' and 'it seems to me . . .' in every sentence to be very intimidating, in the sense that I find what I intended as an offhand comment may have been taken as dogma.

". . . I tend to be oblivious to how others are taking my statements. Many scientists tend to speak in a similar fashion. And the trappings of science, as well, lead to a tendency on the part of nonscientists or scientists-in-training to react as described above."

Intimidation appears as a built-in feature of Jim's field. As a full professor of mathematics and as an observer of his daughters' educations, Jim finds that "all education of math, from grade school to graduate education and beyond, is absolutely sodden with intimidation." He recommends that anyone wishing to see this firsthand, "just go to a few math lectures at a nearby university and count how many times the words 'clearly,' 'obviously,' or 'trivially' are casually used. Now, if you think about it, these words have no linguistic content in terms of conveying information. If something is clear or obvious, you don't need to state it. On the other hand, if it's not, saying that it is isn't going to help, just intimidate. Let me summarize by saying that most teachers at all levels of math are deeply insecure and authoritarian by nature, and when seriously questioned, almost always resort to intimidation."

Jerry has no doubts about scientific intimidation: "You bet your life intimidation is a part of science . . . . I received a review from an 'impartial' NSF reviewer of my postdoctoral fellowship application which went (I would quote exactly if I had it, but this is really close) 'The candidate would do well to serve a period of indentured servitude under a senior disciplinarian who would hold the gun to his head and ensure that the candidate did 100% completed science.' So, a little bondage and domination, a little murder, nice review, huh? Intimidation is not just a part of the culture, is not just acceptable, but it is encouraged. A noted (National Academy of Science member) researcher at my PhD school told a black woman that, being a black woman, she was incapable of doing science and should leave . . . . I'm telling you, there are a million stories . . . . I am sure that the women respondents [to your survey] could regale you for months."

The intimidation discussion brought in more colorful responses than any other. If my small sampling of scientists is any indication of reality, there's an active cult of intimidation in science today. I've included the story of the next respondent because he makes a stab at understanding how intimidating behavior is learned and passed on from one generation of scientists to the next. It's an amusing but telling takeoff on the lively discussion about dysfunctional families being held in the United States today.

Frank, a physicist, learned early in his scientific career about intimidation: "One of my first personal experiences with intimidation occurred about 3 weeks after I joined a research group, in my second year of graduate school. I was asked to give a talk on superconductivity at a joint group meeting. Attendees at this meeting included my advisor, Professor X, five to six of his other grad students, several postdocs and staff scientists, and a similar group of people working under Professor Y. I was very naive and inexperienced, and the talk I prepared was rather shallow, to put it nicely. About 5 minutes into my talk, Professor Y started asking me questions completely over my head . . . . My advisor, Professor X, started yelling at him. As I stood there in front of everyone, probably with my mouth hanging open, the two professors had the academic equivalent of a marital dispute. Seems they were forced into working together because of the large grant they had received jointly, but hated each other's guts. And I was playing the role of the confused child in a custody battle. I kept a very low profile after this for nearly half a year. By that time, I had figured out what was really going on, and realized I could completely ignore Professor Y, as I had my academic parent, Professor X, to protect me.

"At least I had a safe harbor. And although Professor Y was an ass to other people, 'his' people could do no wrong and were well treated. Some people are not as lucky. Several grad students I know are in a much worse situation. Their advisor is very intimidating to her own graduate students. Individually and as a group, the grad students are intelligent, knowledgable, and competent. Each can more than hold their own with a group of peers. However, as soon as their advisor even walks into the room, several of their personalities completely change. They slouch, roll their shoulders inward, and mumble quietly. It's like watching someone assume a completely different personality. This intimidating advisor had an intimidating advisor herself when she was a graduate student.

"This is one of the saddest things about intimidation in science. The abuse and intimidation is passed down as acceptable behavior to the next generation of scientists. The parallels to child abusers who were themselves abused as children is disheartening."

Frank closes his thoughts by pointing out the veracity found in the comedy of radio character Dr. Science, who claims that scientists launch "arrogance attacks" on each other, lobbing their egos back and forth in a never-ending match.

We can learn two things from these responses. The first is that the experience of intimidation in science is not universal. As Jeanine noted above, one can have a scientific training and work experience that is very supportive, very professional, and quite rewarding on an interpersonal level. On the other hand, as both women and men respondents noted, intimidation is common in science, and its negative effects ripple through labs, conferences, and classrooms, leaving its recipients (victims?) appalled, frustrated, and angry; hardly the right mindsets to be encouraged in productive students and professionals. We don't really know if more women than men find intimidation unacceptable, but we can be reasonably sure that it's a scientific attribute that women, who have generally been socialized not to put others down, would be more than happy to see disallowed.

Intuition

Cooperation, competition, and intimidation are all ways of working with others, and their use or misuse affects the social aspect of scientific inquiry. The next characteristic that I'll address is intuition, which is less a way of working than a way of knowing. My reason for discussing intuition is that it has been popularly ascribed as a trait far more developed among women than among men. If this were true, what would women bring to science that men have not? Does science benefit from an intuitive outlook?

Dr. Linda Jean Shepherd, author of *Lifting the Veil: The Feminine Face of Science*,[12] writes of the difference between scientists who primarily employ one of two modes of perception in their work: sensation or intuition. According to Dr. Shepherd:

> In science, the sensation type is the experimentalist who is guided by the facts and is careful not to extrapolate beyond them. . . . While the sensate perceives the details, the intuitive looks for patterns. Without intuition, researchers competently gather data to fill in the holes or increase precision to the next decimal place, but rarely produce anything new. The intuitive is quick to formulate global, sweeping views of problems and generate a large number of interesting hypotheses. But in the extreme, the hypotheses may be fantasies based on no data at all.

It might be construed from Dr. Shepherd's description that it takes two types of scientists to tango: the intuitive to imagine possibilities and the sensate to gather the data. Many scientists probably have a good measure of each characteristic or the field wouldn't have looked interesting to them in the first place. If indeed, as observers have stated, women have cornered the market on intuition, then it follows that women would bring to science a more global perspective, replete with breakthrough ideas and concepts. One of my respondents claimed that intuition and feeling were traits once considered frivolous (and of course, the domain of women) by scientists, but that they are now considered assets, and that women have actually been told that they lack these "essential" scientific skills. Londa Schiebinger points out that it's reckless to embrace "feminine" values as "(necessarily) superior to masculine values," a practice that she thinks "reverses patterns of domination without challenging that domination itself."[13] Is intuition the domain of women? Is it an important part of scientific inquiry? Again, let's hear what scientists have to say.

Carla, the medical researcher, is certain that intuition contributes to scientific success, but she laughs at the way scientists hide the fact, as if intuition is a dirty secret: "A big part of science is hiding how a discovery was made. I do it. Everyone paints a logical scenario for how they were motivated and came to solve a problem. Bullshit. You flop and flail around, and then a flash of insight repeatedly hits you in the noggin, and then you know. . . . My most important discoveries are motivated by feelings of incompleteness in traditional theories and approaches."

Intuition as a way of knowing is critical to the work of some scientists, who have no other way of determining the condition of their subjects. According to Jeanine, a physiologist involved with animal research, "I don't think that, by and large, good results are obtained from animals that are in pain or frightened or generally freaked out. But how does one tell if the animals are behaving normally, are calm, are comfortable? You could call this 'intuition.' It is of course (probably like all so-called intuition) based upon internalized experiences (prior data). And certainly researchers who are men likewise must use their intuition to guess whether the animals are comfortable or not. [Intuition is] *not* a 'feminine' property! [It is] a necessity for anyone who does research using freely behaving animals."

Charles is a theoretical physicist who lays claim to intuition. He is adamant that, "Without intuition, most scientific discoveries wouldn't happen. As a graduate student I studied theoretical physics. One might argue that the mathematics involved would give intuition absolutely no role. On the contrary, when you're doing something that no one else has done before, you'd better be good at making guesses about the proper way to attack the problem. If you aren't, you can spend months or years grinding away without much to show for it.

"I used to solve problems in the shower in the morning. I'd stand and let the hot water run down my back and on occasion I'd get inspired with a possible solution. I'd never see all the details, but I'd get a feeling for how I should proceed. Most often it would turn out to be the correct direction to follow."

John makes disparaging reference to a tenet of difference feminists in his response: "I think it is obvious that intuition plays a large role in scientific research as in most human endeavors. The conversation of researchers is full of statements like 'that looks wrong' or 'it seems to have the right shape.' I believe that intuition is just a subliminalization of experience. That is recognized by most scientists, and so the 'feelings' of experienced people are trusted by themselves and others. It seems to me that few people since the 1950s seriously think that intuition is a feminine quality. The only such statements that I recall are from people claiming it is one of the signs that women are superior to men."

Intuition in science can be hidden behind more traditional methods. Imagine if abstracts to scientific papers included remarks about the actual spark that led to the work at hand. Steve, a 30-year-old astrophysicist at a federal lab, agrees that intuitive thinking is prevalent and yet is not acknowledged in any formal sense: "Good scientists, male and female, use intuition and feeling in their pursuit of science. I believe that these subconscious thought processes (along with plain old luck) are what lead to new insights and breakthroughs. However, intuition, luck, and gut reactions are not approved scientific methods, so when discoveries are reported in the scientific literature, authors rationalize their findings in terms of more accepted approaches."

My final respondent is on record and pulls together a view of the larger academic discussion about sex roles in the sciences. Steven Orzack, an evolutionary biologist at the University of Chicago, has strong ideas about the issue: "Intuition and feeling may or may not be 'feminine' traits, but the real issue is whether female scientists use these qualities to a different degree than do male scientists. To be frank, I think the claim that female scientists use these traits more so than male scientists is preposterous and a self-conscious product of some critics of science who have very shaky evidence in this regard. I must mention that I am, nonetheless, very sympathetic to the intention of many of these critics as long as it is leads to an expansion of what it means to be a scientist. . . . For too long the scientific establishment has deified the paradigmatic scientist who sacrifices everything for their work and who is 'above' subjectivity. This stereotype hurts women *and* men. In any case, an essential

component of the deification of the 'perfect' scientist is that he (not she!) is
above intuition and feeling precisely because these are taken to be unscientific
qualities that prevent one from interpreting nature in an unbiased way (as if
this is possible!). As should be obvious then, the downplaying of intuition and
feeling is very much part of the attempt in the last 100 years to make science a
'profession' as opposed to an avocation (and is also part of the program
promoting scientific 'realism'). What better way to increase the chance that
there will be public support for science (especially financial support) than to
make it seem that scientists are not subject to the same foibles and 'faults' as
are people on the street."

As my respondents pointed out, intuition has always been used in science, and it is certainly becoming more publicly acceptable as a scientific tool.
Because of this, women are not being allowed to "own" the trait as they once
used to. One might respond that men's claims to intuition and other previously defined "women's" traits are meant to assure their continued dominance
in society. I think instead that this is a natural and positive outcome to the
outdated stratification of traits that has for so long kept women out of fields
like science. Even though men are now embracing behaviors they may once
have considered inferior or useless, they are also being forced to consider that
women have the same right to claim behaviors once considered solely masculine. The scientific enterprise could only benefit if it opened itself up to
people who did not fit the established mold of "scientist," including individuals who thrive in openly cooperative groups, people who respect the work of
others and advance their own right to enjoy a respectful and intimidation-free
professional culture, and people who open themselves up to a variety of ways
of knowing, including intuititively.

Women and Ethics

Women will change science, no doubt, by bringing to the field a distinctly different culture. In this section I will briefly examine the possibility
that women may actually change the outcomes or products of science due to
their supposed "differences." Here, the differences that we'll look at are not so
much in style but in substance: the ways in which women may differ from
men in their approach to moral issues, risk assessment, and basic scientific
outlook.

I became keenly aware that women consider ethics in their science when

I received an electronic mail message from Karen, a Canadian forest geneticist with 12 years' experience and a half million dollars of research funding under her belt. She relayed that she had just returned several thousand dollars to the Canadian science funding agency, NSERC, because she had developed a philosophical objection to the work she had proposed, part of which involved "designing 'super' seedlings genetically so that we could cut down all our trees for pulp and paper." Karen "suddenly looked at the big picture and developed an allergy." How many scientific women like Karen are experiencing a shift in what they find acceptable, and how could this be compared to men?

A key theme from some of my respondents was the difference between what a "scientist" does and what, say, an engineer does. When I asked what scientists thought about science that could cause pain, e.g., vivisection, weapons production, toxic chemical production, and so forth, Jeanine, a physiologist, coyly noted that, "the pursuit of basic knowledge should go ahead; the application should be up to the people and their elected representatives. Production of environmentally hazardous materials sounds like, I don't know, chemical engineering or something, not science." Besides invoking the tongue-in-cheek science hierarchy (physicists are better than chemists, chemists are better than engineers, etc.), Jeanine was clear that it was not up to the scientist to control potential applications of their work, but to regulators.

Not all women agreed with her. A female chemical engineer, surprisingly enough, takes the position that, "Because women are, in general, more holistic in their approach to science, the questions they ask tend to be a bit different. Yes, what I'm saying is that, in general, women scientists are different than men scientists. In my job, I have found that my approach to environmental cleanup, policy interpretation, communication with the public, and priorities are affected by my "holistic" approach to science.

"As one example, in chemical engineering school we were taught how to design treatment systems so that effluent concentrations were *just under* the legal limit because this maximizes short-term profit for the company. There was no discussion (except for the profit issue), just equations. If I were the instructor of this course, I would supplement my lecture on design of treatment systems with a discussion on environmental ethics, short-term versus long-term profits, and the obligation we scientists have to consider the effect our actions have on the earth. And I would be considered radical."

Let's take this discussion one step further and examine scientists' involvement in a field fraught with moral ambiguity, where the acts of "regulators" on scientific achievements result in human life or death: the weapons industry.

Gender and the Tools of Destruction

In order to carry out more effective assaults, or threats of assault, scientists and engineers have refined the tools of war to terrifying ends. Early conflict was accompanied by hand-held weapons, with combatants directly engaged in the battle. Today's arsenal makes possible a warfare that depersonalizes the killing and destruction. We've developed weapons to rip the flesh from anyone who treads in the wrong spot, to indiscriminately sear the lungs of inhabitants of an entire town, and to burn down vast tracts of buildings. Our nuclear capability allows us to destroy enormous areas, entire nations really, while poisoning any survivors and the land that sustains them for hundreds of generations.

The great bulk of technological research into weapons development has happened in the last 100 years. Firearms and cannons gave way to increasingly sophisticated explosives, bombs, deadly gasses, and transport systems during World War I. Women, being excluded from powerful political positions, had little to do with the engagement of war. Without their participation at the policy level, it's difficult to say whether women would have been more or less reluctant than men to fund increasingly lethal weapons research. One would imagine, if women truly were the gentler sex, that they would abhor weapons and war. Since in recent history women have not been warriors, can we assume that they have excluded themselves from imagining and building the tools of war?

Not at all. Physicists Caroline Herzenberg and Ruth Howes conducted some research into the participation of women scientists during the creation of the first atomic bomb, and their findings revealed a group of very talented and committed women scientists at all phases of bomb production. In their article, "Women in Weapons Development: The Manhattan Project," Herzenberg and Howes[14] unearthed the histories of about 70 women who worked during World War II in labs at Los Alamos and Oak Ridge, at the Hanford nuclear reservation, and at the University of Chicago and Columbia University. While these women played no part in project direction, they were distinctly aware of the intent of the project, a fact that would seem to diminish any blanket assumptions about women's pacifist natures. So compelling was their work, that physicist Leona Woods Marshall "hid her first pregnancy under overalls and a denim jacket," in order to continue her activities, and physicist Elizabeth Riddle Graves timed her labor contractions before birth with a stopwatch in the laboratory, where she was attempting to complete a set of experiments.[15]

After the bomb was successfully developed and its fury unleashed on Hiroshima and Nagasaki, not all of the Manhattan Project's women scientists were content with the outcomes of their work. Herzenberg and Howes describe how physicist Joan Hinton, who had worked with Enrico Fermi to build an enriched uranium reactor, became "disgusted by what she perceived to be the militarization of physics," became a high-profile peace activist, and eventually left the field and the country to design dairy farms in China.[16] From World War II on, physics and much of science did indeed become harnessed to military uses, with the great bulk of research funding supporting projects that furthered war readiness in one way or another. Many women entered science and became directly involved in the defense industry—with no qualms at all. The end of the Cold War has meant the end of at least part of the colossal US war machine, and with it many of the scientific jobs it supported.

Women, Science, and Environmental Degradation

In addition to forging the tools of human destruction, we might also examine each gender's participation in the modern technological onslaught that has resulted in a threatened planet. Science's contribution to environmental degradation is loaded with personal, professional, and political significance, all factors in sexual "difference." The fact that their work may in some way harm our environment presents scientists with difficult moral and professional dilemmas.

Jane Rissler, Senior Staff Scientist in Biotechnology and Agriculture at the Union of Concerned Scientists, received her PhD in plant pathology from Cornell University. After a short time in academia, Dr. Rissler has spent the past 11 years as a scientist working in the public interest sector, first at the US Environmental Protection Agency, followed by a position at the National Wildlife Federation. In an interview, I asked her whether the potentially destructive applications of science, such as pesticides, nuclear power generation, and weapons, represented what have been traditionally labeled "male" values. She agreed. To illustrate her point, Dr. Rissler recalled a talk by a woman environmentalist who interpreted men's attitudes as scientists in conjunction with their lack of participation in the routine chores of housekeeping. Why would a nuclear engineer who had never changed a child's diaper or had to deal with household waste give much consideration to the byproducts of nuclear fission? According to Dr. Rissler: "I just can't imagine that a group of women scientists, given how women are socialized these days, would ever have developed and promoted a technology like nuclear technology, which we have no idea what to do with the waste and which is so dangerous. To me, I see this business of responsibility, and I almost see it in a sociobiological sense, that men in fact can abdicate responsibility for family and children; and that women in fact cannot."

Again, we see the picture of scientists ensconced in their laboratories, blissfully unaware of the uses being made of their research, blithely looking to the next data point or reaction that will result in another paper. As scientific fields become more fragmented, the kinds of interdisciplinary discussions that might lead to a questioning of the environmental impacts or political ramifications of their work go unspoken. Rissler saves some choice criticism for biotechnology, one of the fastest growing scientific fields, and one into which women are being heavily recruited: "[Biotechnology] is the ultimate in the reductionist approach to looking at living things . . . the idea that we can take things apart and somehow put them back together again better. That is the focus of much of Western biological science. At the same time, we have neglected the more holistic view of nature that says that the whole is really bigger than just the sum of the parts. I see the biotech folks as a culmination of this reductionist view; this arrogance that says that humans can take nature apart and put it back together better. The themes here, this failure to accept responsibility—you just hear it over and over again in the biotech community, 'Well, we just think there are no risks, that we should go ahead.' Doesn't it remind you of the silicone breast controversy? 'Well, silicone is inert, we don't think anything is going to happen, let's don't get the data, let's go ahead.' It's a failure to look beyond an immediate surrounding."

In making her point, Dr. Rissler emphasizes that scientists feel little accountability for the outcomes of their work. In fields like biotechnology, this attitude is particularly dangerous, since scientists' current abilities to forecast the macroscale outcomes of such subfields as genetic engineering are lacking. Would women be more likely to consider the long view in research where outcomes are unclear? This idea will be explored in later chapters during interviews with scientists.

Women, as parents and consumers, have a significant stake in protecting the health of their children from perceived technological harm. It is not clear, from the interviews I conducted, that all women scientists are concerned about the potential risks to public health resulting from their work. Among the women I spoke to or corresponded with, there were eloquent voices on both sides. On the one hand, women believed that all scientific research entailed risk, and that to curtail basic inquiry for fear of adverse outcomes would hamper valuable progress. On the other hand, some respondents were very clear that their science was strongly informed by ethical and practical motivations, and that to embark on research with dubious outcomes, no matter how secure the funding or intriguing the question, was entirely outside the realm of their personal moral code. Certainly the topic is getting quite a lot of attention. In 1990, a membership survey by the American Association for the Advancement of Science indicated that members believed that the "development and articulation of ethical principles" was the most urgent requirement of the science profession.[17] The consideration of scientific ethics was the topic at a 1994 conference convened by the Center for the Study of Ethics in the Professions at the Illinois Institute of Technology. There, a nationwide group of scientists, engineers, and philosophers began exploring how to integrate ethics not just into enrichment classes, but into the fabric of their core courses.

Given that there are variations among the sexes, why might some women still have such strong beliefs about what constitutes value and acceptability in science? One way to understand this may be to observe women's experience of technology in rural Third World communities. In many cultures, sex roles fall along carefully delineated traditions: men are in charge of providing shelter and siring children, and women provide food, water, fuel, and child rearing. Women do their work given the resources readily available to them, whether or not they have an outside income. Land, animals, forests, and bodies of water combine to make up a rich subsistence reservoir. The health of these natural attributes is critically important to their usefulness as raw material. A stream polluted by a factory may kill edible fish; a forest torched by cattle ranchers no longer provides medicines, food, or fuel. In this way, women who survive on a subsistence basis are keenly attuned to nature. The technologies they need to survive are rudimentary, but the technologies that developed countries are imposing on them are complex and often entirely superfluous.

In the conclusion to her book, *Ecofeminism*, Maria Mies[18] shows how differently men and women view modern technology-based economic development in their Third World countries:

> . . . the new vision of a non-exploitative, non-colonial, non-patriarchal society which respects, not destroys nature, did not emanate from research

institutes, UN-organizations or governments, but from grassroots movements, in both the South and the North, who fought and fight for survival. And in these movements it is women who more than men understand that a subsistence perspective is the only guarantee of the survival of all, even the poorest, and not integration into and continuation of the industrial growth system.

Perhaps women who have entered the sciences retain vestiges of the "subsistence perspective" that Mies discusses, and this is what makes up the substance of our difference from men. Women's residual connections to the primal functions that make survival possible for ourselves and our children may provide us with an understanding of what truly constitutes importance on a grand scale, with such comparatively trivial concepts as economic development and technological progress kept in proper perspective. Do enough women retain enough connection to the primal that they will begin to steer science away from some of the technologically interesting, but frequently inadequate and even inhumane ends that it seems so adept at seeking? We're only at the beginning of women's history in science, but we can surely hope so.

4

Transforming the Scientific Workplace

Refer to *Walden Two* by B. F. Skinner for my ideal workplace! It does not, of course, exist. I would like less of a distinction between "work" and "home" and work and play. I would like to put in my share of research and teaching in astronomy, my share of cleaning the community floors and pruning the community garden. I would also like to swim, sing in the choir, take my child and his friends on a hike, all without feeling that I need to get on top of all these "work things." I would like a noncompetitive environment, where we do what we do if, when, and because we like it, without having to consider getting higher, and increasing the value of our resumes.

36-year-old visiting scientist, Harvard University

Science as a Profession: From Pastime to Obsession

In the preceding chapters we've looked at the kinds of characteristics that make for a successful scientist in Western science culture: strictly reductionist, focused on a narrow specialty, very competitive and highly concerned with publishing output. Such a model does not allow much room for obligations or interests outside the laboratory. The model has significant implications for women scientists in particular. In order to become successful scientists, women are expected to give up enormous amounts of what have historically been important parts of their lives: family, community involvement, and other aspects of what were known as "domestic arts."

On the other hand, men in Western cultures have for centuries had society's sanction for playing down their involvement on the domestic front and specializing in work outside the home. This arrangement has been around since the time when, for some much-disputed reason, men left the domicile to hunt and women stayed behind to nurture children. We could fill a book with references to the explanations of anthropologists and historians regarding just why this assignment of roles occurred along sex lines, but for most societies it has been the norm for many years.

Present-day women are by choice or necessity leaving the home and entering the working world in droves. This means women are now facing the same workplace pressures and constraints that have dogged men for a couple of centuries. When science was the domain of clerics and the monied classes, it could be seen as a pleasant avocation; one into which its practitioners could happily lose themselves, mostly unencumbered by the need to bring home a paycheck.[1] During the seventeenth and eighteenth centuries, science was discussed in Parisian salons, laboratories were established in the homes of the curious well-to-do, and findings were self-published and self-distributed. The historical heirs of these leisure scientists are tenured professors. Their combination of comfortable salaries, near-total job security, flexible schedules, and the support structure provided by colleges and universities gives them the ability to pursue their vocation in relative comfort, unfettered by fears of job loss or drastic drop in lifestyle. Admittedly, this level of comfort is not true across the board, but when compared to scientists working on soft money, such as lecturers or lab assistants, the life of a tenured professor is quite appealing. The goal of most academic scientists, then, is to win a tenure track appointment and then sail through the tenure process. An increased competition for grants and a glut of scientists in some fields has justifiably fueled fears among the scientific community that good, tenured positions will be few and far between in the coming years.

There is no position truly comparable to tenured professorship in industry or other private sector work. The economic recession of the early 1990s underscored the precariousness of all professional workers and managers, as thousands lost jobs across the country. So, while a scientist might climb the corporate ladder, perhaps eventually leaving the lab for the executive lounge, she or he is not protected from either the vagaries of the economy or a poorly run company. Though not often publicly acknowledged by today's corporations, who are jumping on the team-based management bandwagon, scientists must compete within their organizations against scientific and nonscientific personnel alike for advancement and job stability.

Scientists no longer have the freedom accorded their wealthy forebears. Instead, they're positioning themselves within their institutions and corporations to get the best-paying, most secure jobs while still pursuing the work they love. What does this mean for those entering scientific careers and especially for women, relative newcomers as they are to the field?

For scientists in industry, it means playing the corporate game, from wearing the proper clothing (suits and hose; a far cry from the relative informality of the graduate lab) to adhering to corporate work hours. Those in academic or government labs have a wider range of clothing options and their work days may be more flexible, but they share one overriding commonality with corporate workers: she who spends the most time at work wins.

The expectation that hours in the lab equal dedication, skill, commitment, professionalism, and all manner of desirable traits came up frequently in my discussions with women scientists. It's not unfair to say that it was the one characteristic about their work that caused them the most angst. I must stress that it was not the act of putting in long hours that troubled these women, but the fact that they had to trade off hours spent in the lab for hours spent with their children and partners. They also expressed sadness that outside interests, such as community involvement, hobbies, or plain old leisure time, were either minimized or completely unattainable. These, however, were far less important to the women than their families, probably because many women scientists, like men, enjoy their work so much that it has achieved hobby status in their lives. To underscore the point: I heard again and again from women scientists who had no qualms whatsoever about their time commitment at the lab or office. As Carol, a computer engineer seeking a tenure-track faculty position puts it, "There are certain things I want to accomplish. Doing that work takes a certain amount of time, whether or not it's a part of my official work week. I do think I would enjoy the freedom of not having to get dressed up and go to my workplace as many hours per week, and I think scheduling things like exercise would become easier, but I'm sure I would still be working 50–60 hours per week. I enjoy and derive great satisfaction from my work. It's not something I seek to escape."

Here is the dilemma: Scientists love science and enjoy working on their research, but they, like most humans, find value in their families and personal lives. (For the purposes of this chapter, I'll assume that many scientists will continue to marry or partner, and that many will have children.) They are competing with others for job security, and their corporation or their academic culture pits them against one another to see who spends the most time on the job. To some degree this has been the male experience for years. But now, instead of men toiling outside of the home while women kept the home fires burning, everyone is at work and no one has enough time to stoke the oven, much less put on tea.

As many of my male survey respondents noted, these pressures are not restricted to women scientists. Though they share a common gender with those who perpetuate academic and corporate cultures, men are frequently caught in a system which they despise. Jacob, an associate professor of seismology, related an anecdote about a world-renowned researcher who demanded the life blood of his students: "One of this professor's students was riding his bicycle near the university, exercising to control his weight problem," he said. "Suddenly the professor pulls aside in his car, and motions for the student to stop. Then the professor demands to know why the student is not in the lab working on any of several important projects. The student protests that he has been working 80 hours a week and needed a short break to clear his head and try to lose some weight. The professor goes into a tirade, saying 'You are a scientist! It does not matter if you are fat!' or something to that effect." Jacob could see the inanity in this exchange, and told me that he was now working to balance his own career with his personal life, which included a professional wife and a 1-year-old child, all the while having more fun.

Men, especially those who are single or childless, may find that they can turn their backs on the relentless pressures of the scientific schedule, or at least can find some time to pursue personal interests for lack of having an involved home life. Men who are financially supporting a family may find breaking out of the scientific routine more threatening to their well-being than staying the course, however difficult they find it. Rod, a 30-year-old on his second post-doc at a large university, suggested that perhaps the reality of "working" so many hours is just a badge of honor: "Not everyone works 70 hours per week. In fact, I would argue that no one does. . . . Speaking from my experience in synthetic organic chemistry (some time ago), I will say that many scientists do indeed put in long hours away from home, but much of that time is spent 'baby-sitting' an experiment to make sure nothing goes wrong, or to be there in case something does go wrong. In theoretical science we spend more intense hours working, but our baby-sitting can usually be done from home." Rod explained to me that he has definite qualms about starting a family given the career demands he feels, and that solutions such as on-site child care would relieve some of the barriers he faces toward being a good parent.

In our culture, however, it is usually women who are expected to baby-sit both the experiment and the family. Considering the time demands placed on working scientists, a natural assumption might be that home life for unassisted women scientists is in danger of falling apart. Scientists can be found among those elite workers who have essentially abandoned their families, communities, and nonscience-related personal development—not to mention fun— all for their careers. The veneer of success may be there: a professional position, a partner, some children, a home, a car, and an annual vacation. But what of the rest of life? What of an opportunity to develop an avocation, to volunteer at a fundraiser, to see a movie? Corporate scientists occasionally mentor high schoolers as on-the-job philanthropy, but perhaps they'd rather be spending the time reading to their preschooler.

Most of the young women scientists interviewed for this book are having an extremely difficult time reconciling what they know to be their future—a tough scramble for a shrinking pool of tenure-track jobs and research funding —with their personal aspirations, which may include parenting and outside interests. Yet their love of science and the commitment they've made to their educations make leaving the field practically unthinkable.

Vicky, a married astronomer who we met in a previous chapter, explained, "I am a graduate student now, and I spend 30 hours alone in classes, meetings, and teaching. That's aside from research and homework. In order to learn in a timely fashion, graduate school could never be a 30-hour-a-week proposition. Unfortunately, I figure my work weeks are about 70 hours. Then I commute 3 hours a day. Yes, it *can* be done!"

Irena, at age 26 is pursuing her PhD in physics, was unusual in that she would even consider the notion of leaving her field for her family: "My biggest fear is that when I have children I might fall out of the loop, so to speak, and that people won't take me seriously. But I'm willing to give it up. I think family is more important than physics."

Women, Children, Science Careers, and the Backlash: Who's To Blame and What's To Fix

Political conservatives have blamed all manner of society's ills on working women. English sociologists Pamela Abbott and Claire Wallace[2] hold that conservatives are reacting negatively to the values of a "permissive society," their term for the sweeping societal changes brought about during the 1960s and 1970s. To conservatives, the "patriarchal nuclear family is seen as natural and universal and other family forms are seen as deviant and immoral." Women who have left the home, found work, and earned an income find themselves emancipated from their dependence on a husband for economic survival. In conservatives' view, working women's deliverance from their traditional roles of husbands' helpmates and children's caretakers threatens to topple the position of men as heads of households, thus ruining the stability of the nuclear family, and by extension the health of all society, since the nuclear family forms the conservatives' acknowledged societal foundation.

Conservatives are not the only ones concerned about women's new roles. The backlash against women for their participation in work outside of the home and for women's emancipation generally is strong across the board and is notable among men. In her book, *Backlash: The Undeclared War against American Women*, Pulitzer prize-winning journalist Susan Faludi[3] exquisitely details the backlash, which manifests itself in all sorts of distorted media and cultural messages. Faludi notes that whatever women's gains have been in the workplace, men's attitudes toward women's emancipation have been less than supportive, because even though "men in the 1980s continued to give lip service to such abstract matters of 'fair play' as the right to equal pay," modern men still want to hold onto their old roles of providers for the family. Faludi points out that women's gains in employment took place against a backdrop of men losing their own footing—with both blue- and white-collar male workers suffering hard times in the 1980s. She provides this insight while considering findings from the American Male Opinion Index, a poll taken in 1988 for *Gentlemen's Quarterly* magazine. Men were quite clear in their preference to remain heads of household, which looks more like a desperate attempt to hold onto their jobs and sense self-worth than to deny women full rights. One way to free up jobs is to keep women out of the market: 60% of the 3000 men polled said wives with small children should stay at home.[4]

Blaming working women for our problems is an easy and far-too-simplistic reaction to a rapidly evolving world. Most women are not fleeing the home in search of wealth; they are working to support their families. The US Department of Commerce census data show that women's incomes play a crucial role in their families' economies, with married women's earnings making up 50% of black family income, 40% of Hispanic family income, and 35% of white family income.[5] To astute observers it is clear that if any group is responsible for leading the nation into its current quagmire of social degradation, it's not working women. Rather, at least part of the blame must go to corporations, their stockholders, and other institutions that establish and perpetuate the kinds of working environments that we've come to accept as "normal" in the United States. Most workers have expectations of putting in 40-hour workweeks, in addition to the time taken for a daily commute, in order to earn enough money to support themselves and their families. What is more normal for many scientists is a 50- to 80-hour week, compounded by frequent out-of-town trips for fieldwork and conferences. There is an unwritten expectation among scientists that the way science should properly be done involves endless hours of toil by heroic (or at the very least stoic) individuals.

When men were the only scientific workers, and were either single with few family obligations or married with an unemployed wife, this kind of schedule may have been tenable. As discussed previously, hundreds of years of cultural acceptance of men being out of the home softened the blow of "losing" one's family to one's employment. Even then, men's personal lives were supported by the women who made sure their clothes were clean, that dinner was on the table, and that the children made it to softball and back. Men do not "lose" their hobbies and free time as working women do, because women continue to bear the brunt of housework and other domestic chores even when both partners work. A 1987–88 survey conducted by *American Demographics* showed that of married men with a child under age 5, "only 30% of the fathers with working wives did three or more hours of daily childcare, compared with 74% of employed married mothers."[6]

Married women have little support for expecting their spouses to share household duties or for asking outsiders to raise their children. Despite nearly 25 years of feminist awareness, women still bear the load of housework, child care, and all the other home-based duties such as looking after aging parents, baking the field-trip cookies, and trimming the shrubs. Men who participate in traditional homemaking activities on top of their outside work, even on a limited basis, have been coined "Super Dads." It's a life that is so rife with stress and burnout that one can scarcely blame men for avoiding it.

Yet women who "have it all" seem to be keeping up, albeit precariously.

In a study conducted with 73 women and 43 men scientists, sociologists Jonathan R. Cole and Harriet Zuckerman[7] noted, interestingly enough, that married women scientists who have children are in fact as productive as their childless counterparts, if productivity can be measured by numbers of publications. The authors point out that this productivity is not without a cost, however. Women scientists with children reported having to "eliminate almost everything but work and family," especially when their children were very young and demanded a great deal of time. This loss of discretionary time, the authors conclude, while not directly affecting their publication rates, may cause other professional difficulties: "Women scientists who adhered to rigid family schedules lost the flexibility to stay late in the laboratory to work on an interesting problem; they report not feeling part of 'the club,' not having time for informal discussions with colleagues."[8]

Corporations by and large have shown little interest in making their employees' personal lives their business. Corporate community involvement has typically meant participation in employee-funded charitable campaigns or employee-staffed volunteer events, whose purposes often reveal self-serving image-building or promotional opportunities for the company. The extent of the relationship between worker and employer is "employment at will," meaning that if workers don't like the conditions found at the workplace, they are not bound to take a job or continue employment there. Scientific workers are usually not members of a union, and they have no collective bargaining power to improve their lot.

Many of the gains made by labor activists were codified by the 1930s, when the National Labor Relations Act was passed. The predominant (but by no means only) structure of the middle- and working-class family at that time was miles away from the modern setup. Mother stayed at home and father was the family breadwinner with a job that often paid enough to support his wife and children. Child-care assistance was sometimes provided by an extended family (as most parents at that time did not migrate great distances from their own parents, as is common today) or by the "family of the street," the neighborhood mothers and older siblings who were on hand while youngsters roamed in relative safety from house to house.

Think about the changes between that time and now: Today, both parents work out of economic necessity, preschool children are in day care, school-age children are latchkey children, and neighborhoods are eerily quiet during business hours. Moving away from one's parents to take a job across the country is common or, in the case of many scientists, a necessity; for to advance in many fields one must accept one or more postdoctoral fellowships away from one's PhD institution, and then move again to a more permanent position that is unlikely to be near where one grew up.

Whereas women used to hold together the fabric of community by applying their talents to volunteer work, the jobs of caring for the poor, looking after our schools, planting trees in the parks, and the myriad other activities that made life more pleasurable now go undone or are left to a few diehard activists who somehow find time to squeeze it into their schedules. Parents are too tired and too guilty to leave their children after work; it was this reality that led to the discussion of "quality time" as a tightly compressed but valuable way to develop relationships with one's mostly absentee children.

What is left of personal and social relationships is unpalatable. Because workers of both sexes are struggling to meet the time expectations of their employers, the amount of discretionary time we have to develop full lives is minimal; a fact that is particularly sad when seen in terms of its effects on families. Without the time to pursue activities of true value, our once-rich culture is being reduced to a parody of work and consumerism. Amy Dacyczyn,[9] a writer on issues of personal economics, points out that the average American consumes twice as many goods and services today as in 1950; that we spend almost three times as much today on hobbies and home recreation; and that even though the average American family today is smaller that in the 1950s, an average new house is twice as large as the average post-World War II house. While essentials such as transportation and housing have become more expensive in real terms, workers today are encouraged to consume nonessential goods and services at an ever-increasing rate, and we do so to satisfy an insatiable emptiness. We are reminded by advertisements that material consumption is the ultimate in success. Our limited leisure time is spent shopping, or consuming sporting events, or maintaining our possessions —washing the car, mowing the lawn, cleaning the house.

The marketplace has often been an important part of successful societies, but never before has it driven an entire culture, across all class levels, to such lengths. The absurd outcome is that Americans are willing to work more hours not just to impress their employer, but to fulfill their ever-inflated consumerist desires. Today's average employed American is working 163 more hours each year than was typical just 20 years ago.[10] Even more appalling, the marketplace tells us that children are an important consumer item, that we deserve them and need to have them; and when we acquire them, we put them away (in day care) when they are an inconvenience to ourselves or our employers. Men and women both have been assured by the all-powerful American marketplace that possessions and purchasing power (ergo: the necessity of gainful employment, even under senseless conditions) are more important than striving for a spiritually rewarding and sustainable existence.

If women accept the charge that their leaving the home is a significant contributor to our social ills, they open themselves up to a backlash and risk losing the personal gains of employment and self-actualization that they'd fought so hard to get. The backlash takes the form of child custody battles, as when Michigan college student Jennifer Ireland was denied custody of her 3-year-old daughter in 1994, because Ireland left the child in day care while attending classes. The presiding judge granted custody to the girl's father, Ireland's former boyfriend, in part because his mother had offered to care for the girl during the day.[11] Women scientists on various computer networks rightfully asked, in response to the Ireland case, what would stop a judge from removing a child from its married parents if both worked and the child grew up in day care? Our society seems desperate to pin the plight of America's children on anything but our obsessive work culture; and since women's entry into the workplace coincided with a steep drop in our country's livability, women bear the brunt of the backlash and the bulk of the blame.

Some of my respondents were adamant that women who made lifestyle choices should be held accountable for them. As Jeanine, a comparative physiology postdoc responding from the Young Scientists Network put it, "I think that in life we make choices, and often those choices carry penalties; perhaps the penalty associated with choosing to have a child is to go nuts from lack of sleep, or to accept less and slower advancement. I see absolutely no reason why child care should fall on the woman's shoulders—it does take two to make a child—and women who want both a career and a child or children ought to choose their partner with great care. To me, fairness means the same standards apply to all; if I choose to work 70 hours per week and as a result I publish more, get more research done, attract more grant money, etc., I believe I *should* be rewarded more than someone who works part-time, publishes less, achieves less, etc. I think that it is not reasonable for people to feel they can have everything without making choices and accepting some penalties."

Jeanine's views were shared by Maria, a computer science professor who made the choice not to have children and who enjoys her combination of successful professional life and outdoor recreational activities. Maria complained that she'd "been annoyed by questions like this for some time," in reference to my query about what a scientist would do with their off-time given a standard 30-hour workweek: "I believe that the premise of this question is just basically at odds with reality. Many top-rate scientists are so engaged in their science that they simply don't want to work on anything else. They are simply going to produce more (probably better) science than many people who are less involved. . . . Many pretty good scientists are also deeply engaged in what they're doing. They will produce more results than scientists of comparable innate ability who don't work as hard. Alas, there will also be people who want to be regarded as top-rate scientists who are driven by that ambition, rather than by the science, and who resent the time it requires."

And so the debate rages. Kindling to stoke the fire is provided by child development expert Penelope Leach[12] and her 1994 book, *Children First*, which, among other prescriptions for healthy parents and their offspring, calls for the end of full-time day care for young children. Leach, with a PhD in psychology from the London School of Economics, is a Fellow of the British Psychological Society and chairwoman of the Child Development Society. She has studied children for more than 25 years, and her 1979 book, *Your Baby and Child: From Birth to Age Five*, has become a parenting bible.

Leach believes that our practice of full-time institutional child rearing is extremely harmful to children, and that the harm it inflicts on the developing child is repaid to society during that child's resulting lifetime of difficulties. Understandably, each time I brought up Leach's assertions to women scientists and science students, I felt like I had ignited a powder keg. A more edgy group could not be found as an audience for such a notion—after all, for a mother to be a student or a professional essentially requires that her preschool and school-aged children attend day care, unless she has access to nannies or a stay-at-home partner. Flames came across the Internet from an older woman scientist and mother who questioned whether I might be accusing women who use day care as, "not following an 'expert's' advice and ruining our children's lives." Almost always, scientists' first reaction was to ask to see "the studies" or to hear about Leach's scientific credentials. These scientists wanted scientific proof, yet objective evidence supporting or refuting Leach's positions has not been published. What we're left with is purely anecdotal evidence. Lack of studies notwithstanding, Leach's convictions have been endorsed by some of the most respected authorities in child development, including Dr. T. Berry Brazelton and Dr. Benjamin Spock.[13]

Leach's premise is that children, especially infants and toddlers, need frequent and consistent attention from their primary caregivers. She doesn't impugn the motives or skills of quality group day-care providers; she simply points out that it is physically impossible for a single caregiver to adequately attend to the needs of more than two infants at a time, and that consistency is impossible in a workforce where high turnover is the norm. Experts and parents divide camps over what "adequate care" consists of, but regulations allow a sole adult to be responsible for at least three and up to eight infants at a time.[14] Leach sees the issue as the importance of acknowledging and responding to an infant's desire to control her caregivers; it is this control that provides the infant with a critical sense of self-esteem, from which the rest of a balanced personality flows. The trouble with day-care kids, Leach thinks, is that they can't get enough attention, and that some of the negative behaviors they've developed relate to securing attention from beleaguered day-care staff—often through aggression, competition, or acting out.

I mentioned Leach's ideas in a question I posted to the electronic discussion boards. They were not well received by Julie, a 35-year-old parent and pathobiologist: "Who in the world is Penelope Leach," she queried, "and how many children did she have in day care while she was trying to 'do' good science? I have had only one child in this situation and positions like Ms. Leach's rip my heart out. The number of hours in day care do not create the child—it's the people, the love, the care, the patience and kindness. Why not 3 hours or 5 or 8—4 hours cannot be supported scientifically. The individual variances are too great!"

Assailing day care is absolutely taboo among working women. The right to accessible and affordable child care is one of the hallmarks of the women's movement, and the strides forward that women have taken, thanks to our ability to get outside help in raising our children, have been enormous. The perennial underavailability of quality, affordable child care has been seen as a major barrier to women's advancement, especially to women who need it to get an education in the first place. Early women scientists often relied on others to raise their children. Marie Curie hired a succession of nurses and servants for her two daughters, and her father-in-law, Dr. Eugene Curie, became their primary caregiver as they grew.[15] Another scientist, astrophysicist Cecilia Payne-Gaposchkin (1900–1979), employed a servant to care for her three children during the 1930s. The outbreak of World War II depleted the stateside workforce as men enlisted and went overseas, resulting in a wide-open and lucrative job market for domestic employees. Payne-Gaposchkin, left without affordable child care, was forced to bring her children to the Harvard Observatory where she and her husband worked, resulting in some predictable difficulties for children, parents, and staff.[16]

A side issue, but one well worth mentioning, is called to mind by the Payne-Gaposchkin dilemma. Their family lost affordable child care when their providers found more lucrative employment. Today, child-care workers' salaries are among the most dismal in the current labor market, where according to a 1994 editorial in *The Christian Science Monitor*, child-care workers routinely earn less than "zoo keepers and parking-lot attendants."[17] A full-time child-care worker in this country often makes little more than the minimum wage; barely one fifth of what a new PhD might expect for her first job after graduation.[18] It's not hard to make the case that child-care workers are being exploited by the educated class: they are asked to be loving, nurturing, instructive, patient, attentive; in fact to replicate the parent in that parent's absence without earning nearly what that parent would expect in the same position. There's an ugly paternalism in the belief that child rearing is a lowly occupation. Despite protests from parents to the contrary, who insist that "it's the most important job in the world," the paternalism surfaces clearly in the wages paid to career child-care workers, some of whom, like nannies, are asked to assimilate into their charges' families. Spending the day with children, day in and day out, can be by turns monotonous and infinitely rewarding; but when done well, its physical, emotional, and intellectual rigors should be recognized as at least as difficult as preparing a paper, lecturing a class, or recording data. Unfortunately, our society has stratified work according to its perceived value, with men's traditional intellectual work winning top marks and women's traditional work of raising children on the bottom rung. Many women who have broken into traditionally male fields are continuing the oppression of other workers.

What is needed is a radical restructuring of the working environments for scientists of both sexes, to accommodate the successful rearing of children, the sustenance of community, and the growth of individuals' moral, physical, and spiritual lives. I must stress *radical*, because what is needed is not just a six-week maternity leave, nor a month-long paternity leave, nor more company-sponsored philanthropic "choices." We will not see much benefit from on-site child-care centers if the child spends 10 hours a day there; from flex-time if it means working from 9 to 7 rather than 8 to 6; or from an academic culture that won't hire it's own, squeezes the lifeblood from those on the tenure track, and wantonly produces so many new PhDs that they'll look forward to a lifetime of professional battles for funding.

As Vicky writes, "Women scientists have to be *very* devoted to their work to climb the ladder. It may be possible to be married (to just the *right* man, and he's not *my* husband!), and to have children (provided her mother lives with her), and have hobbies (they'd better include housework, because you sure won't be able to afford a maid!). Forget community unless you want to give up any free time you ever have. Without very special circumstances, women cannot make it to the top in any field, much less the extremely competitive field of science. I think if you look behind any successful woman, you will see sacrifice that a man would not necessarily know."

Our society has two choices: Either capitulate to the current standard of full-time work and continue to allow families with young children to disintegrate, or develop a new standard for the workplace that provides enough flexibility for parents to excel at their work and at their parenting. Even though several female respondents were adamant that scientists have free will in their choice between children and a career, I doubt they'd want to be thrown out of their own jobs if today's already-conservative political climate took a turn further to the right and gave them no choice at all. By telling a woman that she has to choose between a family and a scientific career, we've shut down a huge percentage of potentially brilliant workers. Must it be this way?

At 48, Sarah describes a life where she made lots of choices and many compromises on her way to her recent tenure-track appointment. As a parent of two and a stepparent of one, Sarah ended up postponing her career: "You have to be willing to do things at different times. When my children were small and at home, I carted them wherever they needed to go, cooked their meals, and helped them with their homework, etc. At the same time, I was active in my church. After the older two left home, I took up karate and made it to third brown before having to give it up due to increased responsibilities with my new job. Now the kids have left home and I have my first tenure-track position. I can devote long hours to work and only have my husband to attend to . . . . I am a firm believer that all things come to those who wait—especially if they are ready for the opportunities when they arise."

But do we have to wait so long? What could Sarah have contributed to her field if she'd been able to do science, perhaps not Nobel-caliber science, but science nonetheless, at the same time she was shuttling her children? And where was her husband during this time? Penelope Leach suggests we look to Sweden for a mentoring system in sustainable work cultures.[19] There, parents of children up to 8 years old can stagger 6-hour workdays, one parent working from 8 AM to 2 PM, for example, and the other parent working from 10 AM to 4 PM, allowing both parents to pursue their careers while necessitating only 4

hours of outside child care each day. These benefits are universal legal entitlements for all workers; employers cannot establish their own "house rules."
Sweden goes on to guarantee either parent 18 months of leave, paid at 90% of
their salary, after the birth of each child.[20] According to Leach, "The difference in attitude is crucial: Swedish children have been given a *right* to their
parents. That's what I want to encourage. A civilized society is not one that
has women in the workplace still bleeding from childbirth."[21]

In her article in the *New York Times Magazine*, writer Gwen Kinkead[22]
suggests that instituting a child-centered social program like Sweden's in the
United States would cost about $175 billion. It would rank with Social
Security and Medicare as one of our nation's biggest entitlements. But as
Kinkead points out, advocates of Leach's program, including Attorney General Janet Reno, argue that "the investment would result in reduced bills for
prisons and welfare in the future." Hardly a difficult choice to make, if one
looked to the long-term outcomes of an all-around positive social program.

What Leach proposes is completely foreign to the realm of the American
psyche. As discussed previously, working long hours has become a badge of
honor in this country, and scientists have refined the concept of long hours to
extraordinary limits. Not only do many scientists not *want* to work fewer
hours, many of them believe that their work would be impossible given less
time on the job. Science, as we've discussed before, is typically a field where
practitioners are driven to seek individual results and personal accolades. For a
scientist to unilaterally shrink her week to 30 hours or less, without changing
the fundamental way science is practiced, would be tantamount to throwing
in the towel, since there are literally thousands of hungry, un- or underemployed scientists waiting in the wings for gainful and rewarding employment, no matter what the time demands. Yet I am thoroughly convinced that
the impetus for change must come from the unified, intelligent, and forceful
actions of women scientists.

So, Where Is This Planet Exactly?

The heading of this section is taken from the first response I received to
an imaginary scenario that I developed and posted to two computer bulletin
boards, the Young Scientists' Network Digest (YSN) and the Women in Science and Engineering Network (WISENET). The purpose of this exercise was to
float various possibilities regarding the scientific workplace across the desks of

scientists and science students and to gauge their responses. I really didn't know how the scenario would play; many of its elements are vastly different from entrenched reality, but are not so unrealistic as to make this a senseless endeavor—especially since some of the scenario's elements were suggested to me by respondents relating the nature of the scientific workplace outside the United States. What is common to each of the scenario's elements is that, during my research for this book, they were mentioned by at least one woman as possible solutions to the difficult scientific work culture found in the United States today. So, in an experiment in futurism, I put them all together. Imagine if this is what the science workplace looked like:

> You are employed in your field, doing science you find engaging. You are paid enough to support yourself, another adult, and two children quite comfortably. You have job security and some hope of advancement, though 'getting to the top' is not terribly important to people in your field, as scientific groups now share responsibilities for most research among colleagues. Tenure has been eliminated. NSF and NIH have greatly increased funding for basic research, following huge public outcry about the waste of American intellectual capital. Your day begins around 9 AM, and routinely ends between 3 or 4 PM. Evening and weekend shifts are given to those who prefer them, but no one is rewarded, either in cash or respect, for putting in extra hours. If the research demands more hours, you hire another scientist. Your group is quite productive and is internationally respected; many scientists want to join it. Its makeup is balanced between men and women. In academic science, group members also divide teaching responsibilities; in fact, teaching is now a pleasurable and honorable task.

The scenario touched a nerve. Responses started coming in immediately from scientists who said they'd been watching my questions on the computer networks but hadn't until now been moved to respond. I asked for and received responses from both women and men, in part to ascertain any differences in their outlook, and in part to see where any commonalities occurred. As we'll see, the scenario prompted responses on polar opposite sides of the coin, and not necessarily along clear gender lines.

After describing the scenario, I asked readers point-blank whether it was desirable. Here's what they said:

Svetlana found the scenario to be the answer to her dreams: "So, where is this planet, exactly? Seriously, I can't imagine any better environment. Of course it's desirable. It leaves you enough time to pursue other interests. It removes the 'do or die' type of pressure that is so prevalent today."

"This sounds like heaven to me," agreed Angela, a 33-year-old university professor of computer science. "No more long looks if I leave at 4 PM, having only been at school for 10 hours. (Especially because I'm single—I've been told that there is no reason for me to go home because I don't have a husband who will complain if I'm not there.) I do not think that my dislike of the pressure and long hours is peculiarly female either. My male counterparts who started with me also resent the demands being made on them."

A third endorsement came from a female physics professor: "To me, this is very desirable. I especially dislike the intense competition that is present in research nowadays. I never knew it would be this way when I decided to study physics. I am planning on a career at a small 4-year college in part so I can avoid some of the competition present at large research universities. My field, particle physics, is especially competitive. I don't mind so much the competition between large research groups, when it's friendly, but I do mind when people in my own experiment steal equipment from me because they can't manage to find any themselves."

On the other side of the coin, I heard from a Chinese male chemist, working in a US lab: "For the best scientists, this sounds like a waste of life. It really takes the zeal away since the life is no different than a programmed robot. There is no thrill, no uncertainty, which is what science is about: discovering the unknowns and experiencing uncertainties by doing so. [This is a desirable scenario] for those who put family in the first place. In fact, for ultra-'family'-orientated [sic] people, it is better to destroy science altogether so people can spend more time loving each other. This may be a male point of view, but I don't know about how women think about science."

Bad news from a female postdoc at a government agency: "Your scenario sounds like the communist euphoria, desirable but unattainable. I say that not because of the dismal employment situation, but simply because the human nature does not measure up to such a scenario."

More incredulity came from a male at a government lab: "The scenario . . . while interesting, is totally unrealistic. What you describe is a technician's job for someone who is just putting in time. People with the ability and drive to do good and meaningful science would feel stifled in this environment. The drive to 'get to the top,' or at least to have your own accomplishments directly recognized by your peers, is a primary motivation for people who undertake difficult and challenging work such as science. No one who performs at an outstanding level in their field does it working only 30 hours a week. That includes scientists, business people, athletes, musicians, and writers."

Last, a male physicist was totally put off: "Heck, if I wanted to get a
9-to-5 job with job security and no homework, I'd become a midlevel bureau-
crat in the federal government! Eighty percent of the 'thrill' of research is in
the competition and the satisfaction of discovery. If you take this away, you
will drive the best, most talented scientists out of science. In fact, what you
have described is a typical job at a big industrial lab . . . . such jobs do *not*
attract the best scientists for the reason that they lack the rivalry of basic
research."

When these replies first flickered across my screen, I had to reread the
scenario to make sure of my wording. The respondents who hated the scenario
were apparently convinced that doing science in cooperative groups would
take away the thrill, the recognition, and the quality of science. Where did this
come from? Is it not enough that the group "is quite productive and is
internationally respected"? Apparently not. What one group of respondents
perceived as a culture and schedule that would free them from undue professional burdens was taken by another group as enforced mediocrity. They look
around their world and see nothing that could improve upon the "science as
personal mission" scenario, where great minds and single-minded laborers
go far.

Of course, this was not a scientific survey, so any conclusions that are
drawn from the respondents are purely interesting and are certainly debatable.
Let's look more specifically at what women liked about the scenario.

Patricia, an engineer, said "The scenario sounds great to me as a future
mother (not any time soon). I think it would benefit society in general if I
could spend some time with my kids and other charities (which I would do). I
also need some digestion time on many of the projects I do, and a full 9-to-5
day sometimes gets so active that there is little time to think about the topic at
a more leisurely pace, and because I want to. I find that the most productive
thought time I have is not frequently when I am at work, but when I am
playing with my puppy. I understand that this may not be typical. I also know
of several engineering departments that have similar lax-time schedules and
many of the profs have consulting jobs on the side. If the prof spends too
much time with his second job, his students may suffer, but there is also the
potential of getting some real-life experience into the classroom—something
that is becoming rare."

Real-life experience? One wonders what a 28-year-old, who has possibly
never held a full-time job outside the academic sphere, who has clocked
incredible numbers hours studying an increasingly precise and specialized

element of her field, and who, because of her downtime deficit, has few if any ties to her community or political surroundings, may have to bring to her work in the way of "real-life experience." What kind of grounding does she have to counsel students on life choices? How can she relate the context of her work or her institution to the larger climate of the outside world? Though Patricia, in her above statement, mentioned it as almost an afterthought, the concept of real-life experience, or well-rounded personalities, or any sort of significant knowledge of the world external to academia or science are certainly downplayed in formal scientific preparation. Luckily for the field, exceptions to this rule always exist; but it's easy to see that in following the dogged trail to scientific achievement, few individuals have the time or inclination to pursue outside interests. The rigors of scientific training leave many young scientists without a clue as to what they'd do if they couldn't be at work:

"I don't really like the idea of working only 5 to 6 hours a day. I love my job and find it hard to leave at the end of the day," says Nell, a computer engineer working for a private company. When pressed, she said, "Again, I'd rather be working, but I guess it would free up a lot of time to develop my hobbies. I'd like to learn to do woodworking, I'd like to learn to fly (become a pilot), and I love to travel, so I'd probably do those things." Nell took issue with the idea that government research funding would be increased, suggesting instead that private industry should see the light and fund more research and development: "Then the amount of money available would have a more realistic dependency: how useful it is, not how much money politicians can steal from people," Nell added.

Cherelle, a population biologist, points out a different difficulty in standardized shortened days. According to her, the possibility of a 30-hour workweek isn't patently offensive, but physically problematic: "This [aspect of the scenario] I can't imagine. My research simply takes time. Imagine a project that required you to be in the field nearly every day for a census, for weeks, or months, on end." She balks at the idea that adding more scientists to the project would eliminate her problem: "Being a field-oriented biologist, I have difficulties imagining parceling out my research to members of a group. Maybe I am competitive, but I suspect I am just a loner."

Joanne, with a PhD in geology close at hand, speaks from the perspective of her own circle of friends, colleagues, and family when she considers the hours' reduction question: "The working schedule you describe reminds me of conditions in Europe. When European researchers come to work at my institution, they often express astonishment at the hours that we keep, and tell us that nobody works on weekends and rarely in evenings in their home institutions, except for a few of the transplanted Americans who apparently like to work long hours. At first I was envious of the Europeans' situation, but only until I began to realize that few of them are doing influential, 'cutting edge' science. It seems that the competitiveness and drive in the US institutions are big factors in how good our science is. I now believe that I would rather be in the US system, because the excitement and challenge of doing really new science is the thing that really makes research worthwhile. Why make the effort if you're just adding data to a stockpile in the literature? What do you gain by surveying yet another rock formation? If I just wanted a 9-to-5 job using my brain in a routine, workmanlike way, I would have become an accountant.

"Also, I would point out that I don't know of any professional career in the United States that does not require long hours. My father is an attorney, my mother is a broker, and many of my college friends have professional careers (lawyers, management consultants, college administrators, etc.). They all work much more than a 40-hour week, and their careers would go down the tubes if they did not . . . . I have concluded that to have an intellectually challenging career that pays above graduate student wages, one has to accept the long work hours. In this sense, I don't feel that scientists have it any harder than other 'professionals.' Perhaps the US society as a whole has gone too far in what it demands of its professionals, but scientists aren't a special class."

Cherelle and Joanne's above statements are typical of other successful scientists and engineers, who have accepted the rules of the game, know how to play, and have taken to the field. For these individuals to think outside of the box is to immediately conjure up the red flag of European scientific mediocrity, the threat of another, possibly inferior, scientist in one's group being allowed to work (encroach?) on one's own experiment, and perhaps the greatest menace of all, the chance that science, done on a part-time basis, would be relegated to an "unprofessional" status. There may be an ingrained fear of the mommy-track backlash in these responses or a justifiable concern that if women fought for and won a more sustainable work schedule, they would be left in the dust (with their children and their unimportant hobbies) while men sprinted off into scientific glory.

Li reports that she and her supervisors are already experiencing a less grueling work schedule, at a well-respected national lab. She says of her bosses: "Their day begins around 8 or 9 AM and ends around 4 or 5 PM (they can leave as early as 3 PM if they need to), but they take half days off here and there, and go on many conferences and trips, even though technically they have 2 weeks of vacation a year. No scientist around here is required to work late or on weekends unless they are working on a proposal or report deadline, but those who work extra hours don't really seem to be rewarded in particular. My boss's group is productive and internationally known and respected—there are visiting scientists here daily from all over the world."

One of the most important statements in the scenario is, "Your group is quite productive and is internationally respected; many scientists want to join it." This sentence was stated as clearly as possible to convey the concept that one needn't necessarily give up prestige, respect, or acclaim if one signed onto a more collaborative research model. Even though it was an intrinsic part of the scenario, many respondents apparently did not see it, did not acknowledge it, or did not accept it as even remotely possible. Their motivations varied. Some felt that their personal competitive styles would be hampered in a group setting; others believed that the highly personal public rewards of the successful lone researcher would be eliminated; still others felt that collaborative groups simply cannot function at the level of traditional hierarchical labs.

Besides the reference to European labs above, several respondents pointed out cultural differences in other countries' science. A woman in industry said: "[Parts of this scenario] used to exist in the communist countries (now ex-communist countries) like Yugoslavia. There was no work pressure there. People didn't work even 30 hours per week. There was job security. The pay was decent. Your hope of advancement was reduced to: 'I have to wait for all these older guys to die (which eventually happens).' The problem was that these scientists were not very much internationally recognized and nobody wanted to join them. But they sure had good lives."

Again, from Cherelle: "The situation I found in Mexico is pretty good. The people tend to work in groups, even the ecologists, usually a principal investigator with a technician and students. Everyone is expected to spend time with family, both children and parents (i.e., one spends weekends, or at least Sundays, taking children to see their grandparents). It used to be (but is no longer true) that salaries were sufficient for a single income to own a house and raise the kids."

Richard, an English physicist, pointed out that, "In France, scientists work much closer to the 9-to-4 hours listed [in the scenario], enjoy 5 to 6 weeks of vacation, and enjoy respectable incomes. There is much more concern about families and children. Day care is provided by employers. Obviously, one pays higher taxes than in the United States.

"Working longer hours is possible and there is always an advantage in publishing first and publishing more. It will help secure a permanent position and will impress the laboratory heads. It may not help your advancement prospects; unless you come from one of the 'Grand Ecoles,' it is virtually impossible to advance into management positions in France. Entrance to these schools is highly competitive and based solely upon a written test. No direct favoritism is possible, the system is fair but quite brutal to those who don't do well on this particular test at age 18.

"I am convinced that there is an intimate link between the more relaxed atmosphere of the workplace [in Europe] and the 'caste' system based upon education. If advancement was more competitive, there would be a stronger incentive to work 'long, hard hours.'"

In addition to the scientists who responded to my scenario, I heard from others who described their experience of the scientific work culture outside the US system. Cherelle, whose views were noted earlier, is a biologist who has spent about equal amounts of her grad school career shuttling between the United States and Costa Rica, Panama, and Mexico. She described a mixed bag of family-friendliness at a large Mexican university: "On days without school, children are around the institution if family care is unavailable. Day-care facilities are more expensive relative to income levels than in the United States, and after 3 PM day care is rare and very expensive. Most scientific women either have at-home private care, or do not come back after din-ner . . . . Remarkably, many [Mexican] men in my generation take turns with their academic wives in child care. To me, the greatest thing is the acceptance of rearing children and the importance of the family. It is, at some level, much easier to manage in Mexico than in the United States, because most everyone views familial obligations as unquestionably more important than immediate research or administrative needs. A sick child requires a parent at home, and either the father or the mother will be there. With younger people, it is more and more difficult to foretell who will go home." Even with this more relaxed attitude toward work, Cherelle reports that the institute she worked with is one of the preeminent groups in the world.

The climate for women scientists in Israel gets glowing reports from Rebecca, a 40-year-old Jewish physicist. Rebecca describes herself as a mother of four, having been "born, raised, and educated" in the United States, then accepting a tenured position in an Israeli university. She reports that Israel is "much, much more family-friendly than the United States. There are objec-tive differences, like universal paid maternity leave and shorter working hours for mothers of young children ([which] doesn't apply to professional staff, unfortunately), as well as a fundamental difference of attitude that has to be experienced to be understood. I can only say that here, professional women don't have to apologize for having families . . . . Since the large majority of women work at least part-time out of the home and use some kind of child care, I have yet to hear the 'don't you feel guilty about leaving your children' thing that one gets with such regularity in the United States." Rebecca points out that many more women study science in Israel than in the United States, with approximately 40% of physics and astronomy students being female. Child care, she says, is "cheap, good, reliable, and stable," with a variety of provider types, including in-home sitters and day-care centers. So stable are child-care arrangements, that according to Rebecca it is not unusual for families to have "had the same babysitter for literally generations." One wonders, after hearing about the Israeli climate for working women, if there is a cause-and-effect relationship between women's high level of persistence in the sciences and their perceptions of being able to pull off both a career and a family life.

Claire, who has dual US and Dutch citizenship, described her experience working as a research chemist at a large oil company in the Netherlands. Employees have flex-time, allowing them about 2 hours leeway for starting and ending work each day. Unlike in most professional scientific workplaces, time cards are used. Overtime must be recorded and is officially tracked by computer. Claire indicates that logging too much overtime is looked upon negatively, as if the employee cannot organize their work or is overwhelmed. Vacation is plentiful, at 27 vacation days per year, and employees are expected to take advantage of vacation time in order to reenergize. Claire also goes on to say that her status as a full-time female worker is unusual in the Netherlands, where only 30% of women work outside the home, and of these women, a mere 5% work full-time. She takes the position that part-time workers are "not serious" about their careers, and is disappointed that, "even younger women are not always allies for working women. Many young women only want to work part-time. And many still stop working altogether at the birth of their first child, even though the percentage of women with college degrees [in the Netherlands] is higher than in the United States. It always surprised me that when I go to women's meetings outside my company, for example, [at] a network for women in science and technology, that more than 75% will say that they want to work part-time or do work part-time. This includes women with managerial positions." Claire closes by pointing out how annoyed Europeans are by the American attitude that science in the United States is of higher quality.

I heard from other women on this subject, who had worked in labs from Australia to Sweden and Germany, each pointing out the relative merits of the non-US systems, and in the case of my respondent in Germany, pointing out the drawbacks, which included academic sexism biased against girls. One of the most interesting responses came from a Canadian professor of applied mathematics, who mentioned that a 27-week maternity leave is guaranteed by Canadian law, funded at two thirds of women's salaries by the federal government. At her particular institution, the federal maternity leave payment is supplemented to the level of 95% for 17 weeks. Part-time tenure-track and tenured positions are also available. The Canadian system appears to be woman- and family-friendly, to say the least.

Of course, we must keep in mind that most of the preceding examples are provided from the perspective of US scientists looking in at the systems as outsiders. It would be fascinating to know how scientists working in other cultures viewed themselves and their relative success (or lack thereof). I have a hunch that such a comparative discussion would reveal vastly different opinions on what constituted success—especially when viewed beyond the commonly used comparison of publishing output. Would an oft-published scientific couple who could not raise their preschoolers, instead, utilizing day care from day one, be considered "successful?" Would a scientist who devoted himself to cutting-edge research, but who lived a social and professional life tightly circumscribed by the scientific community be considered "successful?" Would a woman who was forced to take a second-tier scientific job in order to hold her family life together because her husband would not pull his weight at home be considered "successful?"

I hold this question out for debate. I am not convinced, after hearing dozens of opinions, that there is total acceptance among scientists that what is now considered scientific success is the best model for the field. As Lisa puts it, a successful scientist can have an outside life, "as long as the definitions of a 'successful scientist' change. I believe that as long as one is doing what they enjoy, and at the same time has time enough to spend with her family and enough time for recreation, that is a successful scientist."

Those who stand to gain the most from the traditional notion of success quite expectedly seem to support it, yet even among this group there are doubts. The demise of the halcyon days of full scientific employment is creating a new group of scientific outsiders who went into the field fully expecting to do well, even to excel, by the old rules. It is this group of discontented young people that has to grapple with the validity of current science culture and that has the most to gain through transformation of that culture. Needless to say, they'll have the most to lose if science does not change.

Can science culture change? Two respondents wrote about their current positions, linking some of the progressive elements of the scenario to the realities of their careers. Beatrice, an oceanographer, said: "I received a PhD in 1979 after which I held two postdoctoral positions (1 year each) and then began work here as a 'soft money' researcher in a career-track research scientist position. This means my salary comes solely from my grants that I have submitted to various federal agencies and that have successfully competed and been funded. I worked full-time until I became a parent in 1984, I then took a 6-month leave of absence and have worked (officially) 20 hours a week since that time. Since my child entered first grade I have worked 30 hours a week, sometimes more if a proposal was due or if I was at sea collecting data, but have remained paid at 20 hours a week because more than that would mean more projects and proposals that would require the usual 60 hours a week for a scientist and no leave time for parenting.

"I seem to be successful in that I currently have a 3-year grant from NSF, a grant from NOAA, and a project from the USDA. I hire research assistants and students to work the hours that I cannot. My job category was quite low in status and poorly paid (often assigned to postdocs) until this year. I had the luxury of doing this because my husband has a full-time job with security.

". . . I most often work on projects that involve collaboration among several principal investigators either at my home university or at another. Sharing the credit is usually done by alternating first authorships on papers and presentations at meetings and causes some anxiety but usually only briefly as things are worked out.

"I have actually avoided doing research in very competitive areas that are fraught with arguments and lack of collegiality among investigators, not to mention the time pressures induced by being first out with results. These things distress me so that I refuse to be involved and never have.

"At least for myself and some other female researchers I know, it is not competition or a job with status that motivates me to do good science. It is getting to the answer and doing that correctly, and having the money to get to do it, and enjoying discussions and work with my colleagues. It is very hard to say what those like myself would do if getting funding was even easier and if I had more job security (I do have a security of a sort with [a] spouse's salary). I do not think it would change my work habits much at all. In fact, it sounds like heaven to me."

My final respondent developed a coherent essay on her current work situation as it relates to the scenario. Laurie Mann, a 34-year-old woman with degrees in chemistry and chemical engineering, has experience in both the public and private sectors. In sum, and on record: "My initial response to your scenario is that, in many ways, my present job is similar [to it]. I'm doing applied science (rather than the 'science' of your scenario) that I find engaging. My job is secure. I have flexible hours, a work-at-home policy, good health care, 1-month paid vacation a year, and an on-site fitness center and day-care center. Paternity leave seems to be as common as maternity leave. Leave without pay options are available to most employees who want to take extended vacations. My group is very productive, cooperative, and highly motivated. 'Getting to the top' is not of interest to most of us, as we share a tremendous amount of responsibility in our present positions. We are not paid enough to 'comfortably' support a spouse and children, but it is being done 'uncomfortably' by a few of my co-workers.

"What would I do with more extra time? Do more yard work, more volunteer work, read more, take evening classes in my field, take evening literature classes.

"Obviously this scenario is not impossible—my position with the US Environmental Protection Agency has already integrated many portions of it. To find a job like mine in the private sector, however, would be much more difficult. I don't think that the 'progressive' elements of this scenario need to affect the bottom-line profit margin of private firms in a negative way—I just don't think there are enough people out there running private firms who are interested in making these changes. I've worked for two Fortune 500 engineering firms. Both were very traditional in their demand for 40+ hours per week and almost no vacation. Both companies were run by corporate East Coast offices and very little control (over employee benefits, working conditions, etc.) was given the individual offices or employees. It's very hard to make changes in such an environment.

"My spouse, however, worked for a small consulting firm (one office, 10 people) that was run by a person who loved the work he was doing. He was fairly responsive to people's needs for a personalized environment. Staff-level employees were given a lot of power over their projects and their hours. Extended vacations were no problem. The company made a nice profit. I don't think, however, [that] it's possible to guarantee job security in this kind of environment."

Breaking New Ground: Leaders in New Scientific Working Cultures

Ask any working woman who has worked while pregnant to describe her experience with her employer and it might sound like Laura's, a computer scientist who says, "I work for a white-male-dominated company and am tired of my current pregnancy being such an incredible focal point for anyone who deals with me, especially the various levels of management. I finally decided I don't want to fight these battles anymore. The extent to which the world is easier for white males still astounds me, and it is much more obvious as a pregnant woman!"

Pregnancy and childbirth used to be common reasons for women to leave the workforce. Often it was company policy that a woman lose her job when her pregnancy began to show.[23] With half the workforce now women, corporations have had to rethink their treatment of this normal human function, which in some circles is still considered a disability. After the child is born, Americans are in the unique cultural position of encouraging their mothers to leave them as soon as possible to return to work. The proven benefits of breast-feeding are played down (or made illegal—some states, including New York, Florida, and Virginia, are just now repealing anti-obscenity laws that prohibited public breast-feeding[24]) and the nurturant bond between mother and infant is broken, in the best interests of the corporation.

Things are changing, however. When I spoke with Eileen Prohaska, Director of Licensing at the Kansas City-based STRATCO chemical engineering company, she quite casually described her firm's policy on bringing children to work. Ms. Prohaska, who has a degree in chemical engineering and is a manager at her company, brought her infant son to the office with her for 2½ months. I asked her how she managed to get work done with a baby in tow, and she explained that her son was a "particularly good baby, very content, and there were a lot of other people who were willing to spell me when I needed help."

Mindy Ackerson, STRATCO's Director of Corporate Affairs, described it this way: "We have between 55 and 60 people [at STRATCO] and I can think of one, two, three . . . gosh, we've probably all brought our kids to the office on occasion . . . . It's part of the culture . . . . We have bassinets that just float around, and we have a baby swing that goes from office to office. Everybody accepts it. If you've got to go to a meeting [and your baby is with you], another staff person will hold your baby, or whatever."

At STRATCO, it was president and owner Diane Graham, mother of four, who set the tone of the family-friendly policy by bringing in her own children. Graham has subsequently encouraged other employees to follow suit; allowing staff to extend the period they spend with their infants beyond the standard 2 weeks of paid maternity leave and additional unpaid family leave, plus whatever the employee uses up of their accrued vacation time. Employees also show up at work with their toddlers when they have a day-care snafu or holiday, or when the child may be recovering from an illness but is not yet able to return to day care. A slight drop in productivity is balanced against high returns in employee loyalty and in the overall upbeat tone of the office. At STRATCO, a conscious decision was made to allow employees to mix their personal lives into the corporate culture, and the payoff seems high.

Prohaska also believed that the markedly family-friendly atmosphere at STRATCO made for increased employee retention: "It certainly makes it a lot easier to continue to work. It makes it a lot easier on a working mother to know that if you have a day-care crisis [you can bring your child in], or right after you have a baby you don't have to give it to a stranger at 6 weeks. I worked for Exxon before I came to STRATCO. At the time that I was there, which was about a decade ago, they didn't have any flexible policies on doing anything for mothers when they came back [to work after maternity leave]. Attrition was pretty bad. A lot of women engineers left their jobs. I've heard they have much friendlier policies now."

At a small company like STRATCO, where the management structure is relatively flat and easily responsive to worker needs, developing an employee-centered culture was correspondingly simple. At larger companies, management often refers to the difficulty of applying employee culture changes across the board, where something that might work in one office may not work in a plant. It is up to companies and academic institutions to be creative in figuring out what policies can reasonably apply to individual situations, rather than issuing a blanket ban on progress because of logistical difficulties. I asked Eileen Prohaska if she thought that larger companies could emulate STRAT-CO. She said, "I think it's doable. What you have to look at, I think, is that 2 months is a very small allotment of time out of the number of years an employee will work for you, and if it makes their life that much easier and engenders more loyalty, it's probably worth it in the long run. And you might as well keep someone you might otherwise have lost. When babies are still rather small and are sleeping a lot, you can put them in a swing or cradle. I think you can do that almost anywhere."

How Would Women Change the Workplace?

I believe that the workplace can change. With careful planning, its current tendency toward complete control over our lives and communities can be wrested away. As much as I might find complete devotion to one's work a strange and limiting trait, I would never suggest that humans should be forced to give it up. Rather, the possibilities of a workforce with whole and multi-faceted lives should be experimented with through selected workplace reforms, by which workers could begin to envision a very different but very desirable future. Humans perform best when motivated to do so, not by coercion but by inner drive. Employers and governments need not worry whether the gross national product will plummet when part-time workers are given health bene-fits or when salaried employees are allowed weekends off. In the case of scientists, who are particularly self-motivated, the opportunity to create their own work environment may lead to tremendous productivity and well-being. I've rounded up some representative samples of the observations and sugges-tions made by my respondents regarding useful changes in the scientific workplace.

One respondent who spoke for many others, both male and female, thought institutions should, "Provide easier ways for 'suspension of the clock' in tenure-track positions for childbearing. You see, most of us are now in our 30s when we finally achieve a tenure-track position, and the biological clock is running along with the tenure clock. Any derailment of the tenure clock, by not getting out sufficient publications and maintaining grant support, can cause you to not get tenure. This makes it very, very difficult to start a family."

Reference to the rigidity and petty political squabblings found in some academic settings was made by Carol, a computer programming instructor, "My ideal workplace would have extremely flexible hours, with only attendance at regularly scheduled coordination meetings required. I would actually prefer to work from home most of the time, and I do possess the self-motivation and self-discipline needed to be productive in that circumstance. . . . In an academic environment, I want the freedom to collaborate with researchers and other staff outside my department, without causing a political firestorm. The traditional organization forms (hierarchical and compartmentalized) do not permit this."

Karen, a math professor and single parent to four teenagers, writes that employers could do several things to improve the workplace for women: "Have more of us, for one thing. We have five women among about 95 science, engineering, and mathematics faculty. I am the only woman in a department of 22 mathematicians. It is hard to always stand for all women. Another significant improvement—for parents of both sexes—would be to resist the temptation to schedule all meetings at 4:30 in the afternoon!" Karen also suggested that more flexibility in her department's expectations of working hours would be helpful, noting that, "Senior folks are allowed to work at home, but junior folks have to maintain a presence in the department."

Jerry stood out among male respondents with his thorough examination of excessive working expectations: "I think that the work environment can only improve with family-friendly practices," he said. "The pace that you are expected to keep up at a research university is astounding . . . . At 70 to 80 hours per week for a tenure decision, only those who need no life outside of their field need apply . . . . I will not give up my opportunity for a healthy family—which includes a father the children get to see. Additionally, I might mention that the people who can take these positions are often stunted from the very fact that they must be workaholics in order to gain such a post. While it is possible for a normal well-adjusted individual to attain an academic research post, it is unlikely, as the 'best' candidates have been selected from those who have seen the light of day but once a week, and for maybe 1 week a year for the 5 or 6 years of their PhD, and 2 to 3 years of their postdoctoral research. A change in the workload would allow normal people to enter the ranks."

An astronomy grad student describes her ideal home and work setup: "I would like to do both teaching and research. I think an ideal place is a university, although the politics [there are] not very appealing. I think the normal work week is fine, although I am sure I would bring work home and use a computer link to work at night or weekends. I think a tenured faculty position would be the best job for me. Ideally, day care should be provided by the school. . . . My family (I do plan on having one someday) will be very important to me, and my job will have to accommodate that. I also know that my husband will be supportive and that we will be equals in our household and share responsibilities. . . . I do not think this is a pie-in-the-sky ideal since I know several couples like this."

A heartfelt and disturbing note regarding gender-driven problems in the scientific workplace was written by Carol, who has experience as an electrical and computer engineer in both industry and academia: "I deeply resent the fundamental gender-based inequity of treatment in the workplace. My experiences have taught me that, no matter how competent I am, no matter how much I contribute, I am valued less by my organization than are my white male peers who are less competent and less productive. I receive much lower pay and fewer perks. My ideas receive scant consideration, until they're stolen by a man, who is then given all the credit. On top of everything else, I continue to be routinely subject to sexual harassment. It's more subtle now, usually in the form of ha-ha-wink-wink sexual innuendo, but it's still there and just as virulently demeaning as it ever was. I have also noticed that as my experience and education levels have increased, I have actually been treated less well, encountering greater resistance to my ideas and having my work devalued more. This is opposite [to] what I had expected. Lately, I have come to believe that the underlying cause is that as I progress, I am viewed as a greater threat by both peers and superiors."

Some ground-breaking changes are being experimented with at American universities and corporations. From part-time tenure-track positions, to stop-the-clock clauses, to permanent, part-time professional positions in industry, we are beginning to see signs of the essential workplace revolution. It's important that these experiments receive a fair trial, that they are allowed to evolve to meet the actual needs of participants, and that their existence and successes are widely publicized. Promoters are no doubt going to bear the brunt of ridicule from some quarters; to suggest that an entrenched culture of any sort is deficient is to immediately draw detractors, but to act on the deficiencies is noble at the very least.

The issues raised in this chapter are certainly not the sole domain of women—and as Catherine Didion, executive director of the Association for Women in Science told me, "I hate sometimes when issues get marginalized as 'women's issues,' because then they don't get the due that they deserve. When we hear that family-friendly policies, flex-time, issues of part-time work or reentry of students are seen as women's issues, I can argue that sometimes [the designation] can actually work against some of them."[25]

The scientists quoted in this chapter voiced diametrically opposed views on how the scientific workplace should be structured. All of them are at least partially correct: It is undeniable that those who are motivated to work long hours should garner the rewards for that work. It is equally true that such a work life is not conducive to a well-rounded scientist or a whole individual, much less to a truly healthy family or vital community. Hunter College professor of education Rena Subotnik and Boston College professor Karen Arnold[26] studied the career and personal life trade-offs of 11 elite female scientists in their article, "Passing through the Gates: Career Establishment of Talented Women Scientists." According to the authors, "Science may indeed require single-minded, full-time devotion over an adult lifetime. If this is the case, we ought to discourage family-centered women from aspiring to the top levels of science. However, reconceptualizing the work of scientists has not yet been carried out, and innovative child care and work arrangements point to possible directions for the future." The possibility that women and men might have full lives, as well as satisfying scientific careers, is one well worth pursuing.

II

Science on Women's Terms

5

Youth Leads the Way
Conversations with Young Women in the Sciences

If science or science culture are to undergo any changes because of women's increased participation, we will begin to see them occur with coming generations of young women scientists. The few women who are at the top of their fields today have become so while not significantly rocking the boat of the status quo, in spite of what they went through to break into the profession. The true force of change will come from the next and subsequent generations of scientists; those who have benefited from the women's movement and whose career paths are much less cluttered by the official barriers that hindered their predecessors. Who are these women and what can we expect from them?

This chapter features excerpts of extended interviews conducted over the course of a year with some of science's most forward-thinking newcomers. These scientists were interviewed individually and represent a variety of fields from medicine to engineering. They attend colleges and universities across the country or work at government laboratories, universities, and corporations. We discussed their scientific training, their views on the philosophy, politics, and culture of science, and their experiences with finding and keeping suitable jobs. Their backgrounds, goals, and experiences are quite diverse, but the common themes that arise from these conversations are striking. It is from these interviews that we begin to understand the deep discontent that young women are experiencing in science, as well as their enormous love of their work. In their words one sees eloquence and delight, pride and pleasure, concern, incredulity, humor, and even despair. The observations of these young women on the scientific workplace are startling, and their prescriptions for science merit further discussion. The words of all respondents have been taken verbatim, but names, institutions, and other identifying information have been changed to assure anonymity.

It is my intention to allow the voices of the women in this and the next two chapters to speak directly to the reader. For this reason, I have transcribed portions of interviews directly to the page, with my own commentary limited to the questions I pose. The opinions these women express are particularly interesting taken in context of the concepts described in Chapters 1 through 4; all of the challenges facing women in science are described here in highly personal ways.

Adrianne Bakke is an assistant professor in the biological sciences department at a southwestern university, where she oversees an active research group with three graduate students and three postdocs. She grew up in various university towns in the Midwest, as both of her parents were nonscience academics. Her opinions are informed by her experience conducting research overseas, her involvement in her institution's women's studies department, and her position as a working mother. Dr. Bakke is openly conflicted about the expectations placed on her by her career and the manner in which she would like to raise her young daughter.

Morse Δ Tell me about science.

Bakke Δ Science is a peculiarly insular culture. Very rigid hierarchy. Very male. It's an interesting thing because I think that many women who are scientists did not realize until they were very far along the path that they were doing anything radical. I didn't. I'm 35. That's one of the advantages of being in science—at 35 you're still a young woman!

I have a peculiar appointment. I'm joint between [my department] and the women's studies program. The women's studies program was the driving force behind the appointment because they wanted to get a laboratory bench scientist onto their faculty. Previously they were pretty much just social sciences and humanities; they didn't have anybody who was doing research. And that, I think in part was a reflection in that by the time you are far enough in your scientific career that you have hope of actually establishing yourself, you're not about to, you cannot afford to be political, unless your position is sort of specifically constructed that way, as mine is.

Morse Δ Why can't you afford to be political?

Bakke Δ Well, the structure of research science is—I wouldn't say it's self-satisfied—but there is a sense that we're doing something rather heroic and extraordinary, and to suggest that we're not doing it in perhaps the best possible way is a form of heresy. I like to say that science has become a kind of secular religion, that it is the source of our most meaningful and important explanations for most of the things that happen to everyday people in their lives. Newspaper articles and popular press tend to quote scientists about these kinds of things, and so there's a sense that you're part of the high priesthood and that you must do honor to your profession. And I think most people, most women who are working scientists do honor their profession. We wouldn't be doing it if it wasn't a pleasant pursuit—it demands a lot of you, a lot of personal sacrifices. But at the same time, after a while you begin to recognize that any organization you're a part of has weaknesses and flaws.

Morse Δ What kinds of personal sacrifices do scientists have to make? Especially women?

Bakke Δ Well, I'm sure you must be aware of the structure of the tenure system so the way that things are set up you have to work your ass off while you're a graduate student, that's a given. I know a few people have children during their graduate school years but I don't envy them; it would be an almost impossible task. Then you have to get one or two good postdocs. Increasingly, as tenure-track jobs become more difficult to find, most people do two postdocs, and these postdocs last from a year and a half to 3 years each. So say you have two 2-year postdocs and a 5-year graduate career, so you've been at it pretty seriously for 9 years now. Then you start on the tenure track, and now you *really* have to show your stuff: you need to perform, you need to get your lab organized, you need to begin teaching (probably seriously teaching for the first time in your life, a little task they have not prepared you for in graduate school, because you've been learning how to be a lab scientist), and now boom, you're rewarded. You become a teacher and an administrator, and

you're trying to get money to support your program by writing grant proposals. I wrote four major federal grant proposals last year. At first it's very
difficult to get your feet under you. You're trying to advise your own graduate
students, you're trying to hire and train a technician, you're trying to integrate
yourself into the political life of your department and your university, and if
you have a kid during your pretenure years you're a fool. So there you are with
another 5 or 6 years tacked on, so now we're talking about 15 years, and
assuming you started graduate school when you were 22 or 23, then you're in
your late 30s when you might first *reasonably* begin to think about having a
child.

Well, of course, most of us make unreasonable decisions. I made one and
had a child during my postdoc years. I have a 3-year-old daughter now, but
she doesn't see much of me, and I grieve about that. I mean, it's a source of
personal sorrow. And I also tell myself that I want her to know that all women
are going to work, and that she's going to work when she grows up and have a
meaningful career and the kind of life that I want for her. But I also mourn
the fact that she's not having the kind of connection and affection from me.
Of course I love her and I enjoy spending what time I can with her, but it's
not that much time. And I'm tired.

Morse Δ Are you tired all the time?

Bakke Δ Oh yes. I mean, I think that's a common part of motherhood and an
inevitable part of the early years of the tenure track. I'm in my second year on
the tenure track, and I did two postdocs.

Morse Δ Did you end up geographically relocating?

Bakke Δ Yes, I did my PhD in Colorado, my first postdoc in France, my
second post in Arizona, and then I was hired here. I consider myself extraordi-
narily lucky. My postdoc in France was a delightful experience and a great
exposure to a different way of doing science. There were a number of things
that were different about the French system; the primary one is that, and in
most European countries, big universities just have one professor in each field,
and so there's not the tenure track per se. The lab I worked in had one
professor, a woman as it happened, who was in her mid-50s. It was an all-
female lab apart from the graduate students, not by design, it just had worked
out that way. So there were three women who were the equivalent of advanced
postdocs, really they were in their mid- to late-30s, and they were well along in

their research careers, but they didn't have to do any administration. They worked in the lab and they directed students. They did some teaching and they did a lot of their own research, and because they'd been in the lab from the time they'd been graduate students, they really knew the system in and out. They got along well with each other; there was an intense spirit of cooperation. It took a little getting used to. I remember the first time I went into the lab there, after I'd done one of my first experiments, and the result of the experiment was bacteria either growing or not growing on a plate that I'd left in the incubator all night. As I walked down the hall that morning, I saw the other postdoc, my friend, opening the incubator and taking out *my plate*, and looking at it, and she said, "Oh, it worked." I felt such a sense of invasion that she had found out *first* that my experiment worked. It really ruffled my feathers, and it took me awhile to get used to. I later came to love the sense that everybody knew what I was doing and was following my experiments day to day. It became a real upper to have somebody say, "So what happened when you did that?" or "How did it work yesterday?" It was an unusually intimate connection. I think it's normal that people in the same group exchange information at, say, weekly lab meetings; but this was a much more day-to-day thing, and it was also a source of support when things didn't work, as they often don't. It was a very enriching experience. The other thing I have to say about the French lab is that they were very efficient, they got things done, and nobody worked at night or on weekends except the Chinese students and me. They really valued their families, and they went home.

Morse Δ Was this because it was French culture, or because your colleagues were women, or a little of each?

Bakke Δ I think it was a little of each, but primarily because it was French culture.

Morse Δ Did you take a two-hour lunch break?

Bakke Δ No, and nobody I knew did who was actually working in the lab. I think the people who washed the dishes did and the administrators did, but the people who actually worked in the lab . . . you can't wander away from your experiments for 2 hours in the middle of the day. People would come in at 9 and go home at 7 and that was it. They didn't work nights, and occasionally someone would come in on a weekend to start a culture or something, but they didn't put in a whole day on weekends.

Morse Δ I hear that argument that, "Science is just like that. If you can't work 12-hour days and 6 day weeks, then you shouldn't be a scientist." So—people are actually doing science with reasonable schedules?

Bakke Δ Oh yeah, it's possible, and Europeans in general do that. The British do it, too. It's not that people don't work hard; I suspect it may be that they're more efficient. There's somebody in my department who's famous for always being here. He's supposed to be a terribly hard worker, but in truth one often finds him chatting and having coffee. You can afford to stand around and chat when you know you're going to be there until 10 o'clock at night anyway. There's no rush. But if, by golly, you have to get to day care at 5:30, you do not want to stand around and chat with your colleagues. You want to just get what you have to get done, done.

Morse Δ Is this because men set up the system, and they've generally always had their female support systems at home?

Bakke Δ It's what I refer to as "wifey." The younger men in my department of course don't have wifey anymore, because she doesn't in general exist, but the men in their mid-40s and up have wifey who washes their clothes, makes their lunch, takes care of their children, sees to their airline tickets, you know, every little mundane detail of their lives that the rest of us have to deal with—goes to the dry cleaners—I mean everything—makes cookies for their lab meeting, all the things that it would be nice to have somebody do for you, and it eases your life. But I do think that that's going to change. The only complete workaholics in my department in my generation are unmarried and have no children.

Morse Δ Do they have better reputations? Are they getting ahead faster?

Bakke Δ Well, it's very difficult to say. One of them has an extraordinary reputation and at age 36 is a member of the national academy; he's clearly an unusually gifted scientist in every respect. In part it's because he works very hard and is very organized, and in part because he's brilliant; but he's also not very interested in anything else, and it certainly helps to be completed undistracted. It's just a question of what you want out of your life.

Morse Δ Do you think that at the point where women have parity with men in science, women will say, "All right, that's enough. Now, in the United States, science is going to change"?

Bakke Δ Yes, I suspect they will. And I say this because women in my own department who have children have elected a more reasonable pace of life. And you give something up when you spend more time with your family. To some degree you can offset that by being very bright and very organized and by working hard when you are working. But you can't compete with somebody who's willing to put in 16-hour days, 6 days a week, month in and month out. It depends to some degree on how much your research or your academic progress is dependent on grinding through a bunch of stuff and how much is dependent on thinking of a brilliant experiment or having a brilliant insight. There are people who are not all that bright who put in great long hours and get what they need to get done, done, and then there are people who are extremely bright and can kind of blow in and out. It certainly helps if you are a good leader and inspire your underlings to get going and get lots of stuff done. Then there are the rest of us who fall in the middle ground and who need to put in a good strong 9- or 10-hour day every day to get where we're going, especially when we're burdened about a lot of administrative concerns. But yes, I do think that it will have to change, unless science becomes a kind of refuge for the childless and the unattached, which I don't think is terribly likely. Most scientists I know eventually marry and many of them have children.

Morse Δ Is it possible that if a woman scientist's career took off, she might be able to support a family?

Bakke Δ It would depend on who her partner was. The rather unfortunate thing is that women scientists, entirely out of proportion to what you would expect, marry male scientists. This is not true for male scientists, because there are just not that many of us. It's usually a problem, because there's the problem of getting and keeping jobs in the same town. Assuming you've managed to get that far, then you're both struggling with the same problems. As it happens, I'm involved with somebody who's a full professor and who is 6 or 7 years older than I am. He's in another department. And he's very interested in having a child; but I clearly can't take the time off, and when we sat down and talked about it, I said "I'm willing to go through the pregnancy and lactation, but I can't take any time off now. You will have to take a year's leave of absence to take care of this kid." He thought about it for awhile, and said, "OK, I will." We haven't actually done this yet, so it's just theoretical at this point; but he can afford to do that now, he's 42, and he's got tenure, he's well established in his field. He started in his position when he was 28, which was a very different thing from me, so we're at different stages in our careers. That makes it possible to make these kinds of choices. And I also told him I wouldn't get pregnant unless I got my NSF grant. He wanted to enclose a note to the panel: "Dear sirs, please fund my wife, she wants to have a baby."

Morse Δ When women review other women's proposals, do you think there is any "sisterhood" or other bias?

Bakke Δ Well, I've read various things about this. I know that I have always a bias toward women. When I'm reading a woman's paper, when I'm reading a woman's grant proposal, I am always intensely conscious of the fact that it's a woman. It doesn't always work in her favor, in the sense that I want her to do well and I'm disappointed if she doesn't come up to the standard that I hope for. But I certainly feel much more intense empathy. I think that must be something, I know it's something that African Americans experience, there's just always the sense that you're connected to this person by a common set of experiences. I know that there are also, especially among older, well-established women, a set of attitudes that go something like this: "I made it through this difficult system on my virtues, and by golly you should too." It's not to say that there aren't older women who are very supportive—I certainly have met many—but there are others who are not particularly supportive of women and who in fact may even hold women to a higher standard, unconsciously I think. The women who got their degrees and established themselves in the 1960s must have been the toughest people. I mean it was so hard, and the level of bigotry and the annoying kind of minor problems that they had to deal with because they were women would wear down a weaker soul. So you have to respect the fruits of their experience. It may have been an embittering experience to some degree, but they're still there and they broke the ground for us. It's a complicated problem.

Morse Δ Do women's scientific styles differ from men's? Are we more cooperative?

Bakke Δ It's something that I suspect, but it's really unknowable, and it's not scientific observation. But I still have the sense that there are some differences. I don't attribute these to anything innate—I think they are wholly a product of our different socialization—but I think that it means that women may have a diverse perspective to bring to science, and that should always be welcome. It isn't necessarily always welcome, but in the long run it's going to add fertility to the fields.

Morse △ How did you become a scientist?

Bakke △ I suppose it's because I've always gotten a kind of thrill out of biology. I can still remember getting very excited about seeing for the first time something unseen—seeing the hidden biological mechanism. When I was in eighth grade I had a great teacher who had us study the blood system by looking at the tail of a goldfish under the microscope, wrapped in wet Kleenex or something so it wouldn't die. And as we looked through the tail, which was pretty translucent, you could see the blood vessels and see the blood cells moving through the blood vessels and the valves of the veins opening and closing as the fish's heart beat—everything is kind of golden and glittering under the microscope. It was so vivid and so clear just how it was working, and it was so extraordinary how this suddenly—literally—jumped into focus. It was a breathtaking moment of being overtaken, and I still feel that about my own organism, when I find something new about it, and it's an extraordinary thrill, and it's holy, the way these things work. I think many scientists have a private sense of the spiritual power of biology. It's not something people can talk about without embarrassment usually because spiritual matters are kind of forbidden conversation topics. For some people there's a kind of mystical element to science. I mean, some people are into it just for power or advancement, but at core the initial spark must have been something like the one that I experienced.

Morse △ It certainly couldn't have been the money.

Bakke △ No. Or the glamour! If you were seriously interested in money and glamour, I assume you went to medical school. Most of us could have, and decided not to. There's something about the power that attracted me. Running my own lab is a very satisfying experience. It is certainly rather entrepreneurial, which you discover after you go through the granting process a few times.

Morse △ Are there any pitfalls that women scientists need to be on the lookout for?

Bakke △ There's the teaching track—that is the scientific "mommy track." If you're not careful, women end up doing all the undergraduate teaching and the nonmajors' teaching, which is considered softer teaching. I know that in many other departments, women who have ended up in those areas are seen as more nurturing and more maternal, and also, it is sometimes seen as scutwork, to do the undergrad nonmajors' teaching. However, it's also associated with goodies and status from the university administration at this point because they're very interested in getting these courses on the books; so I think that the administration views it kindly, and our own department doesn't always view it kindly.

There's a danger of letting your consciousness of your gender obsess you. I went to a conference last year on women and science that was sponsored by the Big Ten universities, and I found it to be the most depressing experience. I could barely drag myself out of the place after a couple of days of listening to women talk about what a hard time they were having, and especially listening to women who were associate or full professors who were still struggling. And then hearing all these dreadful stories from undergraduate and graduate students who were in the track and at the end of it and all these depressing statistics from the speakers. At the end of it I thought, "We're doomed. I'll never manage this." And so I think there's a real danger to too much introspection about it. There really does come a point when you have to just say, "I'm a scientist and I'm gonna do my job." Yes, there are these obstacles, and everybody has some kind of obstacle, and I'm not going to obsess about it.

Our second conversation is with Fiona Leighton, who at age 32 is working on her second postdoctoral fellowship in physics. Dr. Leighton received her undergraduate degree from a private women's college and her graduate degree in 1991 from a prominent university, subsequently working in a European lab for her first postdoc. She describes herself as "single, not by choice but because of sacrifice." When she first wrote to me, she was responding to the idealized science scenario discussed in Chapter 4. At that time she said she was seriously considering leaving physics because her internal value system was not reflected in the discipline. She was deeply concerned, mentioning that she "didn't know how much longer [she] could lie to [herself] that things would be OK." I telephoned Dr. Leighton at 9:20 PM, and found her recently arrived home from a full day at the lab. She was eating dinner when I called.

Morse Δ Do you normally eat dinner so late? What kind of schedule do you follow?

Leighton Δ It's not unusual for me to work 60 to 70 hour weeks. On weekends, I probably go in around noon and leave around 5. It depends. I would say I work every other weekend, both Saturday and Sunday, noon to 5. But then, you know, I have a computer at home, and like right now, I just logged in while I was eating, and was just checking on some jobs and stuff, so where do you draw the line?

Morse Δ From your academic history, it looks like you excel in science.

Leighton Δ I've always been good at science, but I've not always gotten good grades in it.

Morse Δ Why not?

Leighton Δ Because I'm a terrible test taker. I really am a terrible test taker, and it's mostly because of fear of failure. I choke. It's very frustrating, when you know that you know the material, you can take tons of practice exams, whatever, and then the real thing comes down and stage fright hits. Actually, during my first semester at college, I got a D in physics.

Morse Δ How did you get over that?

Leighton Δ How did I get over it? I just took another class in physics. I got a B, and then I started getting As, and by the end I was doing an honors project in physics.

Morse Δ Did anybody help you?

Leighton Δ Oh, yeah, they were very supportive there. Again, if I had been at a large state university, I don't know if I would have kept going. But being at a small women's college, they were very keen on keeping me. Although—and this is interesting—the male professors were less interested. The encouragement I got was definitely from the women professors. It's actually pretty clearly demarcated. Isn't that interesting? It's one of those things. For example, the head of the department, who is a male, told me that I was too romantic to have a career in physics. This was already in my junior year, when I was doing well. This guy had a lot more reason to try to discourage me: he was the guy who I'd gotten the D with. The next year, when I came back to sign up for the class, he just kind of looked at me and said, "What? Didn't you have your fill last year?" And I said, "No, no, I'm going to try again." But after I got a B in his class and showed that I knew much better what I was doing, he was very encouraging to me. But the women were always encouraging, and that was good.

Morse Δ Tell me about your view of science.

Leighton Δ I feel that because men have defined what are the valid scientific questions to be asking, then that's defining science. I postulate that women and people from other cultures think differently or have a different viewpoint on life, and therefore science and the questions that they may feel are important to ask of science are not necessarily those that are going to be validated. I think that is one of the problems that science faces if it is going to continue to "allow" women and minorities to become involved in the scientific pursuit. So, opening up the definition of science, reevaluating exactly what it is that science is trying to achieve, might be questions that scientists from the old guard should think about.

For example, one question is the physics of consciousness. I've done a lot of my own informal interviewing around, and I know that a lot of guys think this is a very important question. [These are] people my age and a little bit older than me. I know of at least one extremely well-known physicist at my institution that thinks it's a very good question, but wouldn't be caught dead talking about it. There's an element of heresy now—science has become a religion in which you are not allowed to ask certain types of questions, even if it's just to have an open mind. In other words, you don't want to risk being branded as some New Age kook. But I feel very strongly that physicists, and maybe scientists in general, have become very myopic. This goes along with the fact that we've all become very specialized. As the field has grown and as the technology has grown, it's become less curiosity-driven and sort of more driven by professional solipsism. It's, well, "I can't investigate this question because it was trained out of me, and I won't allow anyone else to investigate it, because apparently it's not what science is. If you want to study the physics of consciousness, go to the religion department."

Morse Δ Is this a male thing? Or is it just that the men who have dominated science set it up this way?

Leighton Δ I'm not willing to say that it's a "male" thing, because I know a lot of men who are extremely interested in these types of questions. I would go for the latter; that it's more because of the way society is a patriarchal system, and physics is a part of that. I think in some sense, physics is taking the male way of thinking to an extreme. And there's great things in that; I mean, every person, every type of person has a lot to offer. I obviously love physics, and there's a lot to be had by this reductionist way of thinking. But I'm not sure if that's the only type of science that can be done. In fact, if I were to go on a gut feeling, I would say that it isn't the only type of science. So therefore, the piece of the universe that we are examining is defined by male terms, and the conclusions are being made in male terms. And so we're only seeing the tip of the iceberg of the universe. Whereas if we allowed various other cultures or women's voices to ask questions that they feel are valid, and still use the scientific method to approach those questions, then I think that we'd be opening up the field in a much broader sense, and I think that it would be valuable for everybody involved.

Morse Δ So who's got the stranglehold on asking the questions? I'm assuming it's the funders.

Leighton Δ Again, I'll refer to solipsism. And this may just be me, I don't know. Maybe there are other people who have more courage, but I definitely feel that I would be branded a nutcase if I seriously went about making a proposal [to study the physics of consciousness]. Forget thinking that I'm going to get funding for it. You never know, right? But certainly I don't think that it is something that the NSF or the DOE have high on their lists. But even before the funding stage, there's a definite bias, or something inherently negative. There's just a fear element there. If I feel that I could set aside some time and write up a proposal to get a grant, I would be staking a lot of my professional career on this. You don't want to do that. That's unfortunate. Then I think, well, I just won't bother asking that kind of question. The risk is too great.

Morse Δ How might that change? What kind of a critical mass of like-minded people would it take before you wouldn't be seen as such an anomaly?

Leighton Δ It depends upon the person. If there were a tenured faculty member, who somehow also managed to keep the respect of his peers, who started to do research into this topic, or other types of topics which were before considered less valid, then obviously all you need is one person. Then there would be people such as I who would choose to work with that person. But of course that's one person, at one institute. I think that's really what it would take. I know another postdoc here, he's a male, he and I have talked a lot about spending a few extra hours a week and trying to set up some stuff in the lab, or try to do this on our own. Eventually, we just sort of look at each other and go, "Yeah, right. Who are we kidding?" We'll just have to put this on the back burner until we get tenure. *If* we do.

Morse Δ So we're not really talking about the feminization of science, we're talking about the humanization of science.

Leighton Δ Yes! For me, that's another problem. Anytime I start talking like this, I get branded as a feminist by some people, and that's not what it's about. Science tends to discriminate against not only women, but just family-oriented men as well! I think it's very unfortunate.

Morse Δ Is the best scientist the one who is one-dimensional?

Leighton Δ I've certainly had discussions about this before. I think the whole point is to be inclusive and not exclusive. That means, yes, of course, if someone really wants to devote their entire life to science, by all means, go right ahead! I would be the last person who wanted to stop someone who has that much passion and drive. That's their decision. But I'd also like the same respect given to me, without feeling less valid as a scientist. Maybe it's unrealistic to think that way, but I think a lot of our problem comes about because of the way our society is defined as well, and the competitive qualities that come out in this type of society. I am one of those people who thinks that the whole society has to change before we can start seeing any solution that has this humanistic quality that we're talking about. Because it's not just science that we see this problem in, it's in the corporate world as well. Until we can have children educated by curiosity and not by competition, I don't think we're going to see a drastic change in any of this type of behavior at all. In other words, where someone isn't judged by how well they perform on a test, but on how compelling their creativity is. The proof is in the pudding. Let a kid be curious about things, and act on that curiosity within reason, and the actions will end up showing the worth of the creativity inherent in that child. It's just extremely clear to me. Unfortunately, people argue that you'd have to change the entire educational system. I'd respond that, yes, that wouldn't be too bad! It's sort of a utopian idea. But I don't know if we can set our sights any other way.

Morse Δ Where can those changes start? Is it public policy? Do we have to legislate this kind of stuff?

Leighton Δ I think it has to be at all different levels at the same time. Sure, public policy, because in the end, politics, as much as I hate it, is the only mechanism we have right now that actually can produce change. For a while it's going to look like forcing something down someone's throat, and that's unfortunate, because you're going to have backlash, you're going to have all this crap coming. I just don't know what the best thing to do is. Certainly, if on global scales, meaning across disciplines and in the corporate world, we can start getting the questions of child care, and parental leave, and all of those types of questions answered in a way in which people don't feel guilty for being family-oriented, then I think that we're off to a good start, within the world of science as well. Just keep on getting as many women as possible into the field. Keep on getting as many minorities as possible into the field. I know that a lot of people argue that affirmative action programs lower the standards, and maybe that's true for a while.

I had an interesting thought the other day. I was riding the bus and I saw this man on a scaffold. He was lifting off part of the scaffolding, and it was just wide enough for him. His arms were just long enough so that he could hold the piece of scaffolding. And I was thinking, "I wonder. The reason that everybody says that women can't do certain things, is because the world was built by men. I wonder if we required all things to be able to be manipulated by women as well as men, then maybe just naturally we'd have women up there doing the things that they need to do." What that would do would be to go to the lowest common denominator of strength. Again, that's something that our animalistic roots go against. Only the strongest survive and that sort of thing. But in the society we're in now, maybe we've evolved to the point where we can take care of the weak. And that it won't kill us as a species to do that. It may end up furthering our spiritual evolution.

I guess, basically, my feeling is that it would take a lot of fundamental changes that would take place within the social structure of our society before changes can take place in any specific field.

Our next conversation is with Susan Ames, a 25-year-old process engineer with a large defense contractor in California. She received her bachelor's degree in physics and has a master's degree in mechanical engineering. She's been in the workforce for 2 years. Several months after we spoke, her employer announced huge layoffs across the board. As of this writing, Ms. Ames is unsure whether her position will be eliminated, and has begun looking for another job while waiting for the news.

Morse Δ Tell me what you think about your company. Have they evolved over the years to become a friendlier place to work?

Ames Δ One of the things that came up the other day is, while my company isn't necessarily on the cutting edge of being an equal opportunity employer as far as women are concerned, they're certainly working hard at it. But we're talking about engineers, who as a lot are very conservative people. There is still the old guard, especially at my company, where the men that are in their 50s and 60s and haven't retired yet, still are very leery of women breaking into the field.

Morse Δ How does that affect you?

Ames Δ Well, in some respects, it's harder to work with them, because I have to work doubly hard to prove myself; but the younger men are a lot more open, and there's a lot of them coming in. You have to prove yourself just like the men around you prove themselves.

Morse Δ Do you think men in the younger generation are more used to working with women?

Ames Δ Yeah, I do.

Morse Δ Does that make it easier for everybody?

Ames Δ The men of the younger generation don't look at me immediately and think "secretary." The older men, when I go to sit in on meetings and stuff, automatically assume I'm a secretary there to take notes, until I prove myself otherwise.

Morse Δ How do you do that?

Ames Δ Well, if questions come up that I have answers to or I have the technical expertise or knowledge, I put it forward and I don't hold back because of their attitude. I just have to stand up and give the knowledge that I have and watch the sometimes-shocked looks on their faces until they rethink their position and quit ignoring me. With some of them it takes longer than others. If there are younger men in the meetings, they will often ask for my opinion, whereas some of the older generation don't even think about it.

Morse Δ Do you think that you have had to be better than most men to get where you are?

Ames Δ I don't think I've had to be better, I think I've had to be a little more outgoing. I've had to stand up and put my ideas forward rather than being asked, whereas in many cases the men will be asked first.

Morse Δ Do you ever get the sense that people might be second-guessing your capabilities because of the emphasis on affirmative action?

Ames Δ Not really at my company. I did at the oil company where I worked before. I got a lot of feeling that I was being second-guessed there. But at my current employer, I have to stand on my own merits or I don't survive, and they know that. I don't feel I'm being second-guessed there.

Morse Δ Are women ever . . . ?

Ames Δ I don't think I've noticed it so much.

Morse Δ Do you consider yourself a feminist?

Ames Δ That depends on your definition of the word. I believe in equality—men and women are just as equal. I don't believe that just because I'm a woman, I should be paid more or less, but I also realize that there are differences between men and women. I'm not about to say that I am a man just to be incredibly equal. There are some differences between us, and I acknowledge those.

Morse Δ Does it seem that once you've put your toe into the wedge of differences, that men, from their position of relative power, can come back and say, for instance, "Well, you acknowledge we're different, therefore you are mandated to take a 2-year maternity leave and not get paid for it"?

Ames Δ I've seen some of that. And they do have some of that hold. If I want a family, I have to take a maternity leave, and while the company I work for now has a paid maternity leave policy, there are some men that I work with that take that as an affront: I get the paid free time off and they don't. And they hold that as leverage—people around me who have taken maternity leave have been held back from promotions due to the time away from the job.

Morse Δ I noticed that at your company it's just a matter of 2-weeks' difference between a 4-week "family leave," used by men, and a woman's 6-week maternity leave—yet women who take the extra 2 weeks are held back from promotions, while men are not. Any chance that this policy would change as more women enter the company?

Ames Δ You know, I really don't know. The hierarchy at my company especially is currently held by literally the old guard. They are changing, but for so many years when they were working, none of this was an issue. I think eventually as more of the younger generation moves up in the ranks it will change, but I don't expect to see it in the next 10 or 20 years.

Morse Δ Will any external pressuring occur, from other companies in the field, that may move some reforms along?

Ames Δ I don't know. What might really have an impact, because we are a defense company, is if President Clinton or the government were to hand down mandates that said maternity leave cannot be held against employees. Because we are so dependent on contracts from them, we would have to comply.

Morse Δ Do you think that would be a good thing?

Ames Δ I think in some respects it would. It's a little unfair, that genetically speaking, women who are in the higher ranks of science and technology are giving up having families or putting them off until extremely late in life, because of the fear of being held back in their careers, or turned down for promotions, or that kind of thing. That's what is holding my husband and myself back—it's the fear that I've only been in this job a year, and it kind of scares me that I'd never be able to get past an E2 level if I had a kid.

Morse Δ And then would you be stuck there until the child entered elementary school?

Ames Δ Probably.

Morse Δ And this is working full-time, too, right?

Ames Δ Yep. The definitions on paper between levels are how much responsibility you can handle and how much supervision you need. Implicit in that is how much time you've spent, how much responsibility you've shown, and how much effort you've put into the job. And that's all manager's discretion, really.

Morse Δ How do you deal with working for a defense contractor?

Ames Δ You know, you kind of have to sit down and say, "This is what I do for a living" and I accept it, and there are reasons for it. Because of my personal beliefs, and that fact that nobody that I work with wants us to ever to have to use what we make, but we all understand that we have to defend ourselves.

Morse Δ Are these defensive weapons only?

Ames Δ There is no such thing as a defensive weapon. All weapons are offensive. The end to which they are used is either offensive or defensive. Every weapon we design is used to go from a gun of some type and hit a target. Whether that target is an incoming missile or tank that's going to attack us, or an enemy base, determines whether it's an offensive or defensive weapon.

Morse △ You mentioned in a survey that the public never asks your male co-workers why they work for a defense contractor, but that they ask you regularly. Do you ever discuss this with your women colleagues?

Ames △ Yeah. We often find that when we do community service outreach and talk to elementary schools and high schools that especially the male teachers, when we start talking about what we do and how we got into being designers, or process engineers, or whatever, the male teachers will come up afterward and say, "Well, what's it like?" We do this in pairs, and the men are also involved, and when I'm making presentations with a man, they always ask me. A lot of the women I've talked to say the same thing. They always come and ask the women who's there, "How can you do this?"

Morse △ Why do you think they do that?

Ames △ I don't know. I think there's this feeling out in the general public that women should never be involved in war, whether that means making the bombs, or working in the army, or going to war, or whatever. There's just this feeling that women are supposed to take care of the children and be very passive, almost, antiwar and peace loving. For a woman to have anything to do with defensive weapons, or weapons of any type, is just almost a foreign concept to most people.

Morse △ Do you ever think about the final outcome of some of the things you're working on? Do you ever visualize the flesh rending and all of that?

Ames △ We tend to think of [the outcomes] more as when we penetrate the target. We're talking about a tank. And yeah, we acknowledge that there are often people in these tanks. But that's one of the things that you really can't think about, if you're gonna do your job. The people who do start to get upset about it find a way out of the company. You don't last if it upsets you that much. We try to avoid the human aspect, because what we are honestly trying to stop is the machines.

Morse △ Tell me about when your company's sexual harassment policy was handed down. Was that in response to a specific situation? Had someone filed a complaint?

Ames △ Oh, man. What happened is, right after all of the government sexual harassment suits—someone came up with one against the President, and there have been several people in Congress—our human resource department decided it would be a good idea to try to implement a company-wide sexual harassment awareness policy. And when they did, they started showing things on the TV monitors around the building, and they started sending memos detailing the sexual harassment policy. Now, the majority of the men I work with are union people, and we had developed a very good working relationship, with the standard jokes—where we each knew our limits—and we had developed those limits over 6 months of working together. And we knew how far we could push and what would not be considered a joke, and as soon as these [memos] started coming down, the whole office, especially around the other young women and I, just got so uptight—you couldn't talk to them any more and you really couldn't exchange the ideas at all. It was almost like they'd put a wall between us. And there was no need for it. It took about 3 months to redevelop that whole series of boundaries on what could and couldn't be said, and what would be taken the right way. It slowed things down for 3 months, where we tried to figure out how to work together again. It scared them, I think.

Morse Δ Could it be that you were in the right place at the wrong time, and that people who are entering the company now will have a more sensitized workforce to work with?

Ames Δ Possibly. I think that every group of people develops its boundaries in a working relationship, and the more sensitive you are to that, is in some ways good, so you don't overstep your bounds early. But to resensitize a group after the boundaries have already been established doesn't do any good.

Morse Δ Did anyone talk with your group to see if this kind of training was necessary? Or could they have created a more individualized program rather than sending out a memo?

Ames Δ They didn't talk to us at all. They overstressed a lot of things. We work around oils and machine dust. If I lean up against a machine and I'm in a nice shirt, the guys would say, "Susan, you need to get this oil off your back," or whatever. And it got to the point where they wouldn't even say that, of even telling me that I'd leaned up against a dirty machine, because one of the things in the policy was "thou shalt not mention anything about other people's clothes." So I walked around with oil all over myself and nobody would tell me.

Dianne Kirchner is pursuing her PhD in chemistry at a prestigious northeastern university. At 25, she is the only African American woman in her department. Her negative experience in graduate school has caused her to re-evaluate her career goals: While she had initially planned to pursue a university position, she is now setting her sights on a lower-stress occupation. I caught up to her after she had written an interesting response about corporate diversity training on the Internet.

Morse Δ How did you get where you are? Why did you become a scientist?

Kirchner Δ Actually, it's because I didn't really like anything else. I know that sounds kind of unusual, but I was always intrigued by science. It was always interesting, and I couldn't imagine myself going into business, and sitting at a desk, and shuffling papers and being bored. If I went into science, I could be creative, I could work with people or by myself if I wanted to, and you know, just work with my hands, which I like to do.

Morse Δ Did you have a good elementary school teacher, or junior high teacher who showed you how interesting science could be, or did you have a role model? How did you first realize that women could do science?

Kirchner Δ Actually, you know, I didn't have any of that. I can't even think of a teacher who influenced me. It was just something I liked. Growing up in my family, my parents taught us that whatever you wanted to do, you could do it, so being a "woman in science" never really crossed my mind—I just liked science. So I decided to go into it. The "woman" issue didn't come into it until later, actually.

Morse Δ How did you decide on chemistry?

Kirchner Δ Well, I didn't like biology very much. It just wasn't as exciting as mixing chemicals and making goo and seeing things change color. Reactions were more rapid, I guess you could say, in chemistry, rather than having to grow something, like cells in biology. You got quicker results, and that was appealing.

Morse Δ Where do you want to go with it?

Kirchner Δ Well, I'm not really sure. I'm kind of in the stage right now, where I'm trying to decide what to do. After seeing the way that a lot of these professors live, and are, and interact with each other, I don't think I want any part of that. I mean, because a lot of it is getting money and publishing. A friend of mine and I were talking the other day about how a lot of professors publish papers about nothing—it's just garbage. But you've got to get papers out there. It just doesn't seem worthwhile. I don't like that about the field. It's very cutthroat in that aspect. And professors—they backstab. It's very political, like any other area, but the underhanded, sneaky things that people do are just incredible! And I didn't know—I had no idea that any of this stuff went on.

Morse Δ How did you find out that this was happening?

Kirchner Δ Just being in the department and seeing how people act. I'm a people-watcher. I really pay attention to how people interact with each other, because I find it interesting. And I would hear things. The department is all white male. I know the way that they treat the secretary, because she's told me they treat her really badly. I hear the jokes that go on. I've heard professors say, "Well, I'm not going to so-and-so's seminar because he didn't go to mine." You know, things like that. People don't share information. There's no communication whatsoever in the department. A lot of the professors don't know how to work with people—they're kind of like in their own environment. They don't know how to handle the grad students, so a lot of people end up leaving, people get discouraged and stuff like that. I'm trying to keep myself away from it, just to survive, but you still see a lot of that going on.

Morse Δ So do you think that academic careers are out for you? Do you think you'll go into industry?

Kirchner Δ I'm thinking about it, but since I have parents who are consultants, I know all about industry, and I know that industry really isn't for me. But I was thinking about that, and if I did go into it, it would be for a short while. I was also thinking about teaching, but in a different environment like a community college, because there I feel I could be more creative and it could be more fun, rather than like the stifling environment here.

Morse Δ If you ended up at a community college, would you be disappointed that you didn't "get to the top"?

Kirchner Δ Well, I figure I can go to the top anywhere I go, but [a community college] is where I think I'll feel more comfortable. I'm not willing to give up my values right now.

Morse Δ From going to scientific meetings you've met other chemistry grad students. Do you think they are finding the same kind of academic culture?

Kirchner Δ Oh, yes. Oh, definitely. With the male students, they don't notice things as much—I mean, they don't have to. But a lot of my female friends are going through the same thing that I'm going through. I have a friend who was thinking about being a professor, and now she doesn't want to do it, because she sees what's going on just like I do, and now she's kind of stuck, because she doesn't know what she's going to do with her life.

Morse Δ Do you think that the whole environment, and the things that you don't like, are going to change when more young people and more women are involved?

Kirchner Δ Yes, but I'm thinking it has to change before that. Because you have to keep women in school first, and that's a problem, especially at my university. I talked to the secretary, and I found out that the department was one third women: 27 women students out of 88. And mostly women drop out. I thought that was incredible. I thought at least it was, maybe, 50/50, but it's not. And more and more women are leaving. In order to change the environment, you have to keep the women there. But there's not a lot there for us in science, because of the kind of environment that it is. It's a sink-or-swim environment. A lot of people struggle, but no one is willing to help you. A lot of people I know have gone to counseling at the campus women's center.

Morse Δ What's it like being African American in that sea of whiteness?

Kirchner Δ The funny thing is, the issue for me is more about being a woman than it is about being African American. I was worried about how my professor would treat me, being an African American and a woman. I'm not having any problems there; he's a very patient person. Other people in the department treat me a certain way because I'm a woman, not because I'm African American. I haven't noticed any blatant racism or anything like that. I think it's kind of unusual that it happened that way. I mean, I'm glad . . . if you think about it, it's OK. I can give you a funny story. My professor wrote a follow-up proposal to NSF, and on one of the pages, he wrote down all of the grad students that worked on a certain project that he was writing about, and next to everyone's name, he put "grad student, grad student, grad student . . .," and then next to mine, he put "Dianne Kirchner, black woman graduate student." And my reaction was the same as yours, I just started laughing, you know? Other than that, no one acknowledges the fact that I'm African American. It's other things.

Morse Δ Do you feel that you stand out in any way?

Kirchner Δ Oh, definitely. I'm the only black female in the chemistry department, and there are three black people total, out of 88. Yes, I stand out, but I don't think that to stand out to people at this university means "outside."

Morse Δ Do you remember any peer pressure? Especially when you were in high school? Did any of your friends influence you in one way or another?

Kirchner Δ No, not really. I kind of kept to myself in high school. I hung around with other smart people—I was in my little clique, as people usually are, in high school. And we all really had different aspirations.

Morse Δ Do you have any science heroines or heroes now?

Kirchner Δ Well, something that I'm proud of is the fact that my grandfather taught chemistry. He was a child prodigy. He went to college at 14, and got his PhD, and he graduated summa cum laude. He was first in his class. This was back during the time when black men just didn't get into college. He was one of the best chemistry teachers at his institution, and that, to me, was very inspiring. I come from a very educated family. Other than that, there are a lot of African-American scientists that I admire, and actually, when I was a TA [teaching assistant] for general chemistry for the freshmen, I used to put up little facts about African-American scientists on the board, just to let them know that there were people who invented things, you know, common things, that they probably didn't know about.

Morse Δ Now that you're in science, how might you change it?

Kirchner Δ Well, I can tell you one thing that I did. I started a liaison committee. My class was pretty large, there were about 42 of us coming in. A lot of people had problems, and there was a chairman, of course, but no one had the final say-so on a lot of things. So people got depressed, and didn't know who to talk to, and didn't know where to get answers, and things like that. So I started this committee, and a group of us just got together and talked about what needed to be changed. I took it to the chairman, and he said he would change a few things.

I'd also change the recruiting procedures. The African-American people that they admit, they expect them to go through this special program where you come in over the summer, and it's only for African-American people; no one else is invited to do this. I got upset about that and raised my voice, and got them to change the process.

Morse Δ What's the point of that program?

Kirchner Δ Just so that students can get acclimated, and work in the lab for a while, which I think is great. But why should we be letting just the black people do it?

Morse Δ Maybe the shy people too?

Kirchner Δ Everybody! Whoever thinks they want to do it, or needs it, or whatever. There are some other people, with low GRE scores, who need to be in there more than some of us who come along. I actually didn't do it, because I had a good job over the summer.

Morse Δ Is part of the awfulness in academia born out of the fact that more people are competing for limited dollars?

Kirchner Δ Yeah, part of it is that. Part of it is the fact that it's male-dominated, and right now I don't want to be the ground breaker. I know some people have to be, but I'd rather break ground in another area, because I think I'd kill somebody. I swear, these people are crazy sometimes. The things they do, it's just incredible.

Morse Δ How old are they?

Kirchner Δ Forties, I'd say. Late 30s and 40s. We have two new young professors who just joined, and as a matter of fact, there's going to be a woman coming. They've been trying to get a woman for years. Now they've finally got someone who's a postdoc. We're going to see what happens when she comes along. It's going to be very hard for her.

Morse Δ So these guys aren't very old; they're just midcareer.

Kirchner Δ Right. Midcareer and clueless. It's something that I'd never expected. I didn't know what to expect, but I didn't expect this.

Morse Δ Do you feel betrayed in any sense?

Kirchner Δ No I don't, because I didn't know what to expect. If I had expected more, or something different, I would feel betrayed. I don't care for it. Right now, I just want to get my degree and get out, and then make a difference elsewhere. Someplace where I can flourish.

Carol Klein is a second-year medical resident at an urban hospital in the Northeast. At 35, she's begun a second career after spending several years as a researcher in microbiology. Dr. Klein's experiences as a medical student and resident have made her bitter about her career choice, and she hopes to add her voice to the growing number of physicians who would seek to redesign medical training.

Morse Δ Why did you become a doctor?

Klein Δ I think I wanted to for a really long time, since I was a little lass. I think nursing was the only thing that I ever saw, so I thought I was going to be a nurse when I was real young. I loved to give shots and stuff, and that was cool. And then, when it came to the reality of high school and college, I didn't think that I could do it. Going to med school wasn't totally at the top of my consciousness. But toward the end of college I started thinking, I *can* do this, I *should* do this! And I changed my whole curriculum about a year and a half before I graduated college, in order to have a premed degree, in case I wanted to go.

Morse Δ Why did you wait?

Klein Δ To sort things out, experiment, see what was going on. I'd heard about research and the first I'd ever heard of it was in college. I didn't know anything about it.

Morse Δ And what kind of work did you do?

Klein Δ I worked in molecular genetic work for 5 years. I did some pharmacology, and I did some clinical GI stuff—gastroenterology stuff, working on the gut. So I did a lot of animal research, and bacteria and virus and DNA studies.

Morse Δ Now that you're in it, what would you change?

Klein Δ Oh, God. I would change the . . . entire way we're trained. We're pummeled. We're exhausted. We're worked so hard, that it's inhuman. More importantly, it's dangerous. I'm talking about after med school, now. I think it's dangerous for the patient that the least experienced doctors, that's me, are making decisions at 3 AM with no sleep. So in the middle of the night it's just fine for us to be that decision maker on the frontlines; but come 8 AM when everybody's bright-eyed in the morning, then it's fine for the attending physicians to stick us back into the subservient, bottom of the totem pole role, because now they're there. But magically, when they go home, we can make those decisions.

Morse Δ So you're being used.

Klein Δ Very much so. I feel used and abused, completely. I mean, we work, when residents say they are on call, what that usually means is going to work extremely early, say sixish. I'll give you an example. I did cardiac surgery last month. I would meet with my higher-level residents as late as possible, so we'd say, "Let's meet at 6:15 AM" instead of 6 AM, so that would mean a big break . . . and we would meet and round on our patients, which means that we hope we don't have a lot of patients. We usually know how many we have, but one or two might have come in during the night. We go around and see what's happened during the night. We examine them, we have to plan the day and write orders and a note about what's going on, like "Stop this med, start that med, transfer them here, check this, do that." Then we'd write to ourselves to remember what the heck it is we were asking for so we could check on it later. And we might have anywhere from 3 to 12 to 17 patients, and we power around . . . sometimes we divide up, the group might only see two people, we don't all see every patient, so at the end you [remind each other], "Now, [regarding] Mrs. Longtree, don't forget to check this, it's vital.""OK, I'll look at that, I hope I'll remember." Then we have to make it to a teaching conference by 7 AM. So there, in 45 minutes, we've seen 15 patients. Then at the conference, we have to pay attention, and maybe get called on, and put on the spot, you know, "Dr. Klein, what would you do if this patient blah, blah, blah?" And then the operating room starts at 8 AM, and you have to be ready to go. So maybe at 10 minutes to 8, you run over to the operating room, find out what room you're in, get your gear on, which is laborious, put on all your stuff, and have all your concentration now be just on that surgery, even though in the back of your mind you've just seen 15 patients, and a couple of them might be really sick. You've got all these things going on, and you hope it just flies without you. And then you concentrate on the surgery. You're often paged during the surgery. And you can be on call, which means you stay all night long, and when people go home, you also take care of their patients. Then you do the same thing the next morning with no sleep.

Morse Δ How many hours straight does this go on?

Klein Δ Thirty-six to 40 hours. Then you come back at 6 o'clock again the next morning. That's how it works—and you're lucky if you have a day off a week.

Morse Δ And this goes on for how many years?

Klein Δ If you're in surgery, 5 years. So I would change that fact that we're abused. I think we could have shift work—why can't you do a 10- or 12-hour shift and hand it off to the next person? We need more people, and that costs money. And residents run hospitals.

Morse Δ So who perpetuates this system?

Klein Δ I think the hospitals that have us as their employees are one component. They get a big lump of money, and I don't know these facts exactly, but out of that lump of say $150,000, we're paid $30,000, and the rest, you know, they make money. They have to keep certain standards up, and the place that I work is a lot better than most; they have to be accredited, they have to teach us certain things, but we are their doctors. We're called the "house staff." It's like the military, only I borrowed $60,000 to join. I'm in debt $60,000 and am working this hard; granted, I've changed my specialty now, I'm not in surgery, and my schedule is better.

Morse Δ Your specialty is radiology?

Klein Δ Right, I just started that portion. I did a transitional year internship in a variety of areas.

Another thing I would change is call. I don't even think call is necessary —and saying that is like blasphemy. "Well, there's no other way to do it." But I don't think that it teaches you anything [but] to be so exhausted that you just cry, and falling asleep and 5 minutes later you're up again for the rest of the night. You just lay down and you're getting paged—your pager is going off every 5 or 10 minutes. It's amazing how your standards just plummet when you're so busy. You have to prioritize. Obviously, if you think someone is going to stop breathing, you don't care if someone's nauseated. You truly do not care. It's not a joke—you don't give a damn. But that nurse has only talked to you about *their* patient. And all they know is, "Dr. Klein is not coming to see the patient. She said you'll be lucky to see her in 3 to 4 hours." And I'll just be honest and say I can't come until then, you can try this, but . . . and I truly don't care. The minute I get off the phone, that's out of my mind, because so-and-so might be dying and I'm not quite sure what I should do yet. That's human nature—everybody would do that. But they put you in that spot all the time.

Morse Δ Do nurses support female interns and residents more than they do men?

Klein Δ I wouldn't say more, no way. Not more. I think you have to prove yourself. I have seen charming men get fawned over, definitely, definitely. And called "doctor" more. Whereas I would more likely be called "Carol."

Morse △ So there are some interesting dynamics happening there.

Klein △ It's very blatant. I think, me personally, I've met a lot of nice friends who are nurses, who treated me more respectfully than most. But only I think because I've worked before, I understood what's happening, and I was considerate. And they knew that. I never, ever was mean to a nurse. Ever.

Morse △ Do you sense in any cases that women who are nurses might have gone into medicine instead, and are somewhat resentful of you?

Klein △ Maybe a little bit. But you see, they know so much, that it's almost like they think, "Oh, God, why did you do it, honey?" In a way they're kind of supportive, in that they'll say, "Look what they're doing to you." They wouldn't do it themselves because they know how it is at a real intimate level, where I didn't really appreciate how bad it was. So it depends. I think if you make it a team thing, people are bound to help you, just like any job, anywhere. It's just hard to do at 3 AM when you've been up for 45 hours. It's hard to keep it up.

Morse △ So what have you given up in your life?

Klein △ I've given up hours of free time that I don't even want to think about. I've given up a sense of freedom. I've given up a lot of friendships and potential friendships, because we don't have time to develop them the way normal people might. We certainly financially have given up plenty—I mean, I borrowed money and didn't work, so combine what I used to earn, multiply it by four years, and add what I had to borrow. When you give up financial leverage, you're really vulnerable. I can't just walk in there and quit today without giving it 6 months of serious thought because I'd have to start paying those loans—$600 a month *at least* in payments—and how could I do that if I worked in a coffee shop? And if I didn't owe money I could say, well, that's OK, I can just quit now.

Morse △ So you're an indentured servant.

Klein △ Very much so.

Morse △ If you had to do this all over again, would you?

Klein △ I think that, sadly, it kills me to admit it—I've always been really positive and optimistic. I know it's a roundabout way of answering, but when

I was going into medical school, and had decided I was going to go and was accepted, many, many doctors said, "Oh, God, don't do it." Or, "I'm sorry." And I just looked at them and said, "I'm not like you." If I'm not happy, I don't keep doing what I'm doing. I hate people who complain about what they do and still do it. I'm not like that. And now I have to bite my tongue and not say the same thing to people who are all excited about going to medical school. I try to say the same thing without being judgmental about it. I say, "Well, maybe you could find out about this or that, make your own decision, but I didn't realize about this or that (say, the hours) and maybe you should hang out with a resident." I still would go to medical school, but I wouldn't go straight into residency. I'd take time. It's a big steamroller process, med school. There are certain deadlines, and criteria and goals that everybody's supposed to have, and really ways you are supposed to feel, in a way; you're supposed to be excited that you're finishing med school; you're supposed to know what you want to do as far as a specialty in your third year of med school, which is way too early. You're supposed to have that decision made, be excited about it, have the ball rolling, and spend thousands more dollars flying around the country interviewing while you're still going to med school. Then they let you know where you're going in March on "match day," where a big computer puts your priority list in along with the priority list of the people who interviewed you, and it spits out the answer about where you're going to live and work.

Morse Δ Would it help if you had fewer patients? If your caseload was smaller?

Klein Δ Oh, sure. Of course. But the reality is that medicine is going in very much the opposite direction. I mean, you're allotted 10, 15 minutes per patient in some clinics.

Morse Δ And with the talk of health care reform and health management organizations it's being tightly regulated.

Klein Δ Yeah, so now we'll have even less control. You're trained to think a certain way, but then you're not allowed to do it. You might as well forget it. So I chose a specialty that's different, so I'm removed from those things. I don't have to follow the patient when I go home. When I go home, I'm done. I'll be a consultant to my colleagues, which I like. I love interacting with colleagues, and talking about the ins and outs of something. But even if it's a horrible thing, I can still go home, or go on that sailing trip at 5 o'clock that I said I'd do, despite the fact that Mrs. Jones now has bleeding in her head. Even though I diagnosed it, someone else has to stay.

As you can see, I'm really disillusioned. Very. And it makes me sad.

Morse Δ Women seem to have been socialized to accept the nurturing, "mommy" role. Are women more nurturing?

Klein Δ It's funny, I guess women are more perceptive. But you don't have a lot of time for it, a lot of it gets shoved aside. I think that I actually am more nurturing and aware of what's cooking than most, and I do it, but I can't keep it up. It takes time, and you really need the time to go sit in front of the computer and go through this person's test results. What I would do is try to do both, and be very tired. Tired, emotionally and physically, because you have to be *on*. I feel like I'm a waitress, "Hi, how are you today?" I mean, you'll walk in and patients will ask you for another blanket. I hate being a female doctor sometimes. There have been times, when instead of getting the blanket, which is my normal human reaction, I'll say, "You push that button for your nurse. I'm your doctor." And I don't like saying it, I'm embarrassed at myself for saying it, but they are like little kids laying up in bed. "I want potato chips now," or "I didn't like that Jell-O today." I'll go in and say, "Remember me, I'm Dr. Klein, how are you feeling today?" and they'll say, "Well, this food!" Basically, I don't want to hear about the food. Then later, I'll get a call from the nurse, "Do you know that Mr. So-and-So has bad stomach pain?" And I'll say that I asked him how he felt that morning, and he said he didn't like the Jell-O.

Morse Δ Do they do that to men?

Klein Δ Oh, somewhat. We've certainly swapped these stories. Sometimes they just don't like to complain to the doctor about the real stuff, and then the minute we leave they tell the nurse and then the nurse will page us. And it pisses the nurses off, because they listen to complaints all day, and then the doctor shows up, and when you ask how they are, they say, "I'm fine! How are you, doctor?"

Morse Δ Here's the big question. And this is a little different, because I've been talking with women who are in scientific fields where women are unbelievably underrepresented. Medicine is not like that—women are entering the field in large numbers. How is medicine affected now that more and more women are becoming physicians?

Klein Δ I think what it's doing is adding more and more frustration to the field. Everybody's frustrated because if we just accept the way it is and join men, then we're all too busy. Everybody's too busy, the male half of the partnership and the female half, whereas before, men went and did their thing, they were the doctors, and they had a happy home life because the little woman was at home doing it all. So the laundry was done, and the kids were fed, and you could go home and actually talk to them. But now, men and women, and it doesn't always follow gender lines, there's a lot of men who feel the way I do too, they want to see their kids. They do *not* want to live like that anymore, and they're mad too. If it doesn't change, we're all going to be really angry and frustrated, because now as women, we want to have children or continue to have lives outside of work, and a lot of men want to be with their kids because they feel that a lot of them were cheated out of their fathers, and they don't want their careers to be *everything* to them. The men that I am friends with have either some significant other or a lot of interests that are really diverse. They're really good at what they do, and the reason they're really good at it is because they keep it in perspective.

It doesn't have to be like this. Sadly, what I see is if it doesn't change, medicine is going to lose people like myself and my male friends too, who walk away from it. And I think that in some of the fields where you need better people, you won't have them, because we've said forget it. We won't put up with it, so we go into other kinds of fields. Life's too precious to get sucked up, and we don't get off on being the all-important doctor that needs to be called in from the Sunday picnic with their family. I want to stay at the family picnic, and so do my friends. I'm going to have a life.

Patrice Mitchell is an award-winning French-Canadian earth scientist at a Canadian university. At 33, she has faced the difficult balancing act of nurturing her science career while bearing and raising a child. With her research grant up for renewal in a few months, Dr. Mitchell is philosophical about her future in academic science, especially since she and her husband would like more children. If her grant is not renewed, Dr. Mitchell will "gratefully look for more flexible, forgiving alternatives," than the academic tenure track. She says, "the current tenure system may force me to choose between academia and a decent family life."

Morse Δ You've been following some of the threads I've been launching on the Internet about women and science—how do you respond to this line of inquiry?

Mitchell Δ I find that it's important for me to respond, especially just having now gotten a start [in my career], and really seeing that collide with the experience of having a baby so early in the process. It has given me the perspective that I knew I was missing. It's like all of a sudden being on the other side of a hill, a divide, in that I felt that throughout my education and graduate studies, I've never really met any open sexism. I did not think that my experience was necessarily exceptional. I figured that some people do run into problems and a certain number don't. And that difference alone, I thought, was the main factor why women eventually reach a ceiling in their careers.

I remember very clearly having a conversation with a male graduate student in the United States where I did my PhD, late at night, and telling him that I felt that the crunch for me was really going to come when I was going to start having a family. I felt I had no model about how to do it. I had a very good idea from my graduate study of the amount of work that it was going to involve; it was a department that was very research oriented, at a relatively high profile in our field, but I felt that I had absolutely no model, or guidance, or any idea of how one would manage the juggling act that had to come with starting a family and still maintaining the high level of performance that is required today to carve a niche in the research community.

Morse Δ What exactly is your field?

Mitchell Δ Earth sciences. My own work is very laboratory oriented. It requires not only a lot of time, but it is something that you cannot do from home. To just get the research started—coming back to Canada at this university—that is what has really been the most difficult. Just to get my research program going I have to spend a lot of time in the lab, and it's something I had to cut back during the pregnancy itself, because there is not an awful lot of reliable information about the risks that are involved in what you are manipulating. There's still not that much known about it. And when you're the only woman alone in the department who is facing those questions, you really do not know exactly where to turn. So I found that I soft-pedaled my research program, probably more than I would have if I had been better informed during my pregnancy.

Upon coming back from my 10-month leave, [my] energy level was very much depressed by the whole experience. There was an added complication. I found after a few weeks of coming back that I was pregnant again. I had a miscarriage two months later. It was just another blow: I was just trying to get back into it, and it just kind of set me back physically again. I had just a little bit more to overcome. Mentally, it was just hard to get through for a few months, because there was not really anyone who could relate to what I was going through. So I feel that it's really that isolation—just the fact of being alone in my position. Even among the other science departments here at my institution, there are no other young women with families right now, trying to do what I'm trying to do. And I find that the biggest source of stress for me is not having any type of blueprint at all, or mentor, that could relate specifically to some of the things I'm going through.

Morse Δ There are certainly young men who are starting families.

Mitchell Δ They are, yes. There is one at the moment in my department, and I can see the indirect effects of the strain on him. I see the course evaluations compiled for him. I know how strong he started, just as I did, and [during] the last two terms [his evaluations have] kind of gone low. If I had to bet any money on what is putting this extra strain on him, I know it is the fatigue that goes with a new addition to the family. They had one young kid, and a new one is not exactly easy.

Morse Δ Has your university provided any child-care assistance or other support?

Mitchell Δ It's very scant. There is a child-care program here, but it accommodates only about 125 children for the entire university community, which actually compares quite well with other universities across Canada. But that still means that Steven has been on the waiting list since before he was born, and it's almost 2 years since then, and there's still no room for him. They make a big effort to give priority to people who are living in even harder situations, single parents or single-parent graduate students, so in that respect the resources are definitely not adequate.

Morse Δ Are you married? Does your husband help out with child care?

Mitchell Δ Yes. He's as involved as he can be within the time that he takes away from work, but that happens not to be a lot. He is working for a consulting firm, and that means that he really cannot dictate his own terms. We have felt that we didn't want to put Steven, before he was almost 2, into a regular child-care arrangement. I am the one who has been accommodating my schedule around what we feel is the best balance. So I find that in terms of time, I am the one who has really gone a long way. Essentially I would say I've put my career on the line. On other planes, I find that every time that [my husband] is back home, he definitely makes Steven his priority. He's very supportive and very involved with him.

Morse Δ Can science be done on a part-time basis? My idea involves a radical restructuring of science practice to a more cooperative, group-oriented activity, where practitioners could spread around some of the demanding hours. Can that happen?

Mitchell Δ I think it can. I've heard really encouraging reports of part-time, tenure-track positions that are being experimented with in New Zealand and also in England, I don't remember if it's at Oxford, but one of the examples in the earth sciences was brought up in the seminar at one of our national meetings a couple of years ago. They seem to be having really promising results. What they did was to prorate the remuneration and also the expectations in terms of productivity. And they found that if you combined the productivity of two people occupying half-time positions, it was often comparable or higher than a single person, because if you're really passionate about what you do, you manage to do a lot in the time that you have, especially if you know that your position is going to have some stability. You can really work, invest for long-term results, and remain productive. That's what I've been able to do in the last 6 to 8 months. I'm doing it part-time because even though I am officially full-time, I'm still accommodating our son's schedule to quite a large extent. I find that I've made very steady progress and it should really blossom this year. And I feel that from then on, in terms of the grant process and so on, if I survive this year I will really be back on track. And I find that I've really been able to do that on what amounts to about a half-time position.

Morse Δ Are you happy?

Mitchell Δ I'm happy that so far I have respected my priorities. I hurt at times that I've had so little reinforcement in my image in the department itself. But I'm glad that so far I have not compromised the things that I really believe in. I don't feel that I've left any students in limbo as grad students. I find that where I have to make compromises, I haven't forgotten where my priorities are.

Morse Δ Have you ever felt that anyone has ever impugned or questioned your abilities because you received a special Natural Sciences and Engineering Research Council (NSERC) award for women?

Mitchell Δ Not so far, not in this department. But I'll tell you that when my PhD supervisor had to be part of the process in recommending me for [the award], he told me that in general he really frowns on that type of program, and that the reason he was ready to participate in this process for me is that he figured I would do well no matter what the competition. And therefore, he didn't feel that he was propping someone, basically, by allowing me to get into that program.

Morse Δ Do you think that's a danger in some cases, that women who might not be as well equipped . . .

Mitchell Δ I find that in the review process that goes on that's really built in, and the fact that faculty positions are so tight as they are, that universities really have to make the choices that will pay off in the long term, in terms of productivity and so on. They have to be so hard-nosed about it. If someone weak, relatively weak, for example, was admitted because of what looks like some type of preferential program, [the person is] just not going to survive the evaluation process. So I find that it's an important door to open. If there had not been any other way of getting into the system, I might have been side-tracked, just because the baby could have arrived at the same time anyway. Now, at least I feel that I have a track record, even in getting the grant. They have to show that they have some kind of commitment to not letting me be derailed just by having had a baby. I'm going to find out whether [the grant program] is actually addressing the issue it's supposed to address or whether it's only window dressing. I don't feel that in any way it could do a disservice.

There are people who talk about the possibility that less than desirable candidates might find their way into the system, but we have such an ongoing review process—every 3 or 4 years you end up being evaluated. Even people who are well established are under such pressure, and get cut for what seem like really subtle ups and downs in their productivity. Someone who is clearly deficient is just not going to be spared.

Morse Δ How are things for scientists in Canada? Are they as difficult as they are for people here in the States?

Mitchell Δ Things are hard here, too. The program of NSERC, of individual research grants, is really under assault right now, and that is making everyone nervous. We feel that it was really one thing that we were really proud of. We could maintain research programs that were really modest and still be sustained. My perception from my stay in the States was that NSF put so much stock on really large projects, you really have to be big in order to just keep going, or to be extremely well connected. I find that the operating grant system here at NSERC makes it basically easier for a first-time applicant to get into the system, and they give you a chance to prove yourself.

Morse △ How would you change science as a woman?

Mitchell △ Right now I think I'm especially sensitized to the impact that childbearing has. I feel that's not going to change much because women are going to tend to always take the most active role in early child raising and so on. I think that, as far as our programs go in Canada, I would not change much, but I would want NSERC to at least articulate better its policy of what a woman can expect in terms of how she's going to be judged when those factors come into play in her career. I've found that even in this program, prior to my maternity leave, when I contacted the director of the program, she could not give me any feeling for how they were going to review my case. It was a long time before I learned [about my review process], because I floated several suggestions about coming back part-time, for a term at least, and then coming back full-time a term later. They were totally at a loss—they couldn't give me any examples of solutions from people who were in my situation—there was just no background that they could provide me. That is really the problem for women. There is a scarcity not only of role models, but the funding agency itself doesn't have clearly set guidelines. And sometimes you need those. That has been a very hard thing to cope with. It's one thing to say you should feel confident enough to set your own terms, but when there's literally no precedent on campus, I have very little to go on.

Morse △ Why should society, or anyone, make adjustments for people who have children?

Mitchell △ Because the people who are active now and are contributing members of society were children once themselves. I feel that if they have a lot to offer, it was because they received a lot as they were growing up. Therefore, if you want people to keep giving back to society, they have to have a good start. And they can't do that if their parents face impossible choices. I think it means that a certain number of parents will be extremely conflicted, some will make choices that they may regret, and the children don't have a voice. We need to have more flexibility. Especially because nowadays there are so many types of families. I feel that it's just investing in the future.

Julie Hegvik, age 33, has just received tenure at a southwestern state university. She holds graduate degrees in chemistry and education. She told me that one of the reasons that she made it through the tenure process so quickly was because she had wanted to have a baby before her biological clock ran out. Dr. Hegvik has studied feminist epistemology, and contacted me to provide her contributions upon learning about this book.

Morse Δ You are 33-years-old and you have received tenure—you're doing very well. How did you get through so quickly?

Hegvik Δ I have always been very determined that that was what I was going to do. I finished up my graduate program pretty quickly. Partly, also, because my husband was ahead of me and I didn't want him to have to drag on [waiting for me] forever. And also just because I felt like I wanted to get done. I guess I've always set pretty ambitious goals for myself. I'm a real long-term thinker, a long-term planner. I knew a long time ago that I'd want to have a family, but that I'd want to be secure before I did. For that reason I very much planned to have a baby after I got tenure, and I knew I needed to do that pretty quickly before the old biological clock ran out.

Morse Δ So you thought about this when you were 22, 23 . . .

Hegvik Δ Well, maybe not quite that young, more like 25, 26 . . . by the time I was in graduate school, I was thinking in those terms.

Morse Δ Let's talk about the feminist critique of science. I've been reading much about difference feminism, and whether women naturally find that they are less competitive and more cooperative, and more peaceful, and more loving, and all of the positive things that some feminists are saying might be ascribed to women as a group. If that is true, does that affect their science?

Hegvik Δ Well, I think that there are some of those qualities that women are perhaps encouraged to take on. But I would certainly never say that they were biological. And I would also say there's huge variation from one individual to the next as to the extent to which they take on those traits. You're talking about socialization, and some people are more socialized into kind of a conforming set of values than others. I guess I would think that there's a wide variation. I guess I would also think that the women who are most likely to have these kind of nurturing, cooperative kinds of attitudes are less likely to go into science. We probably filter those people out pretty quickly in our educational system.

Morse Δ Is that a good thing?

Hegvik Δ No. I think that the major problem with the way that we're educating scientists right now is that we're attracting people with a very narrow set of values and dispositions. I think that's very detrimental to science. But, I think there's kind of this idea about what science is that's very cultural, that's part of our educational system, it's part of our cultural beliefs, and people believe that's the way science is, and that it's incompatible with a lot of the qualities that girls and women are encouraged to take on.

Morse Δ How do you think that the educational system and the culture of science itself might change?

Hegvik Δ Both of these institutions are very hard to change. Schools are *enormously* resistant to change. They only change superficially. I think that teacher education is extremely important along these lines, so that teachers are aware of the implicit message that they're sending their students about what the nature of science is and what that has to do with their gendered selves. I think it's important that teachers examine it very closely. It's one of the reasons why teacher education programs ought to have courses about what science is, because I think that a lot of teachers' views about what science is are largely based on a myth, rather than on any kind of informed view. Or perhaps I should say that a lot of times they take on a view of science that scientists themselves take on. It's not informed by those who study science. It's not informed by views of historians and philosophers of science, who really make it their business to study science at a metalevel. I think the only way that you're ever going to change science is by increasing the spectrum of people and ideas and dispositions that do science. I don't think that's just a matter of getting more women, but getting more men who don't fit the traditional mold; more people of different ethnicities and cultural backgrounds and experiences.

Morse Δ Do you think that science is one of the last bastions of fields without diversity?

Hegvik Δ Perhaps. I'm hesitating, because I'm trying to think of any counter-examples.

Morse Δ I can think of fields such as the trades . . .

Hegvik Δ Yes, like carpentry or plumbing. [When] you look at academia, the most male-dominated areas are engineering and the physical sciences.

Morse Δ Why would women want to go into those fields? I'm being the devil's advocate.

Hegvik Δ I can tell you why I majored in chemistry. The reason I majored in chemistry is not because I ever wanted to be a chemist, but because at the time I thought I might want to go into one of the health professions. It was a very good ticket for that; I knew that getting into a good medical school would be highly enhanced if I had a chemistry degree, especially over a biology degree. I never wanted a PhD in chemistry, because I didn't want to do chemical research. I was only interested in chemistry because it allowed me to do other things. I enjoyed teaching it, but I never wanted to be a chemist. I'm married to a chemist and my father is a chemist.

Morse Δ So you had a lot of support.

Hegvik Δ I was never scared of it. I think that's a reason why a lot of women don't go into chemistry and physics. And not just women, but a lot of people don't go into those fields because they're scared. You know, when you're 18 years old, you're pretty cocky. I figured if Dad could do it, I could do it.

Morse Δ What do you think people's views of science are at age 18, especially if they don't have parental role models who are involved in the sciences? What kinds of popular culture images are informing their decisions?

Hegvik Δ I think [science] is very much seen as reserved for the brainy, if you will. It's for people who are very bright and have a particular way of thinking.

Morse Δ I would assert that you have to be very bright to do many of these things. They're complex. Are people who are bright in other ways not allowed to develop a scientific way of thinking?

Hegvik Δ I guess that's part of it. There are a lot of other subjects that are also complex, yet people don't view them that way. To understand, for example, and to understand well, a lot of literature in sociology and to understand it in important new ways takes very, very bright people. Yet I don't think it has the same kind of mythical status as science does.

Morse Δ How do the sciences create their own mythology?

Hegvik Δ It has such a long history. It goes all the way back to the Enlightenment, when science was really coming into being in Western Europe, that science needed to distance itself from anything that was social and anything that was cultural, largely because of the politics of the time. So they created a science, and a view of science, that was very objective and very detached. That view of science unfortunately stayed with us. Perhaps people want to believe that science is all these things. Perhaps they want to believe that because that gives them a certain amount of security, that there is a way of knowing that gives us the absolute truth that we can count on.

Morse Δ Like a religion.

Hegvik Δ Yes. We like to have experts that can tell us exactly what to do, so we don't need to think about it too much for ourselves.

Morse Δ How does that bear on having women entering the sciences? We generally haven't thought of women as intellectual powerhouses or as people we can trust with important issues. We live in such a paternal culture.

Hegvik Δ It's a case where you have very-high-status knowledge and very-low-status knowers. Men have not thought women particularly capable of doing that, and a lot of women bought into it.

Morse Δ Do you think they still might?

Hegvik Δ Some of them do. Some of them just don't like science—not that they think they're not capable, they just don't want to do it. They'd rather do other things. I teach a lot of elementary education majors, so now I'm working with a group of students who are very different from the kind of students I went to school with as a chemistry major. There's a lot of conviction among my students that they really want to make a difference, and they think that the way that they can do that is through the lives of children.

Morse Δ I've been talking with men about their view of science, and have been requesting their opinions about doing science and simultaneously "having a life." I've gotten several very emphatic responses from some men, as well as from some women, to the effect that because one loves one's science so much, "having a life" is definitely secondary. They don't even bring up having children or becoming a part of their community. That doesn't seem to matter. Might there be an extremely well-suited science personality, and is it good or bad to encourage or discourage this kind of person from spending their lives, literally, in the lab?

Hegvik Δ Well, exactly what is it they're going to turn out?

Morse Δ Data! Information! These people swear to me up and down that they're doing vitally important science, and it's up to politicians to set the public policy as to what happens with their findings.

Hegvik Δ Well, that's a very nice kind of illusion to live under. It's funny, because when I was in graduate school, when I first started really reading a lot about science, and the history and philosophy of science, I talked to my husband about it. I remember the night I read [Thomas S.] Kuhn's *The Structure of Scientific Revolutions*. I was describing this to my husband, and it thoroughly shook the ground that he stood on. He was so upset, because he really wanted to believe that what he was doing was the absolute search for pure knowledge, and that it was the best thing you could do with your time because it was the search for the truth. It was very upsetting to him when we started talking about some of these issues, and he finally accepted that what he was doing wasn't all that different from what a lot of people were doing—that it was not a search for pure knowledge, that there were a lot of other things going on as well.

Morse Δ Can you put into your own words what Kuhn was saying those "other things" were?

Hegvik Δ He talked a lot about how scientists work within paradigms. So much of what they do is heavily influenced by what was done before them. The kinds of things that are already believed are going to heavily influence in what direction you take your research and what kinds of capabilities you have as far as instrumentation. What you actually observe and how you interpret it are going to be influenced by the paradigm within which you work, and so you're never able to quite escape the strong influence of history. Scientists are very much working with the development of historical knowledge. Because the sciences are so highly theoretical in that sense, what knowledge is produced is always influenced by what's gone on before, so you can't really make the claim that the work that you're doing is pure and untainted by any other prior beliefs or expectations.

Having some sort of an awareness of what you're doing, so that you're not just doing research without being aware . . . I guess my beef with a lot of scientists is that they have no critical self-awareness; they just do what they do, and there's no reflection on it. Perhaps if they were educated a bit differently, perhaps they would develop the capacity to think critically about the work that they do. Maybe that would make a difference. If scientists were even a little bit aware of the history of their own discipline, then maybe they wouldn't be so naive. I don't think these are bad people we're talking about. I would hope that they would be a little more careful about what kinds of projects they undertook and things of that nature; if they took responsibility for the outcomes of what they do.

Morse Δ Can women help that along?

Hegvik Δ I think so. Now, whether they're going to do it any better than men, I don't know. The problem is, if you take a woman and educate her the same way you educate a man, you're going to get the same results. I don't think that just getting a few women in is just going to make a difference, if you don't change the way they're educated. I think that women are as capable of producing bad science as men are.

Emily Winthrop is an angry woman. She has had an almost unbelievable set of experiences in her short science career, from sexual harassment to stalkings. It is possible that her attempts at publishing research have been blackballed by her peers in retribution for work she published on the difficulties of women in her field. She is a 31-year-old assistant professor in an agricultural science at a southern university, who has decided to resign her tenure-track position with no other job waiting, because in her words, she "can't take it any longer."

Morse Δ You were on the tenure track.

Winthrop Δ Yep. I was tenure track. Three hundred people applied for my job, and I got it. That's how bad the job market is. That's pretty common; you'll hear that from anybody. I've kept my applications out there since the year before I got my PhD. From 1990 to this year, I've been applying for jobs. And everywhere I go, it's been at least a hundred applicants. Even in my limited field. It's very bad.

Morse Δ Tell me about your work.

Winthrop Δ I know that whether women do different research from men is one of your interests, and I think they do. My research has really been shaped by my feminist interests, because what I'm looking at is essentially chemical ecology. Female [insects] emit a chemical that attracts male [insects], and people have known this since the late 1950s. Chemists have found ways to make these chemical pheromones and synthetically release them, and the idea is that you can confuse the poor males; they think there are females everywhere, so they fly around and they don't ever find the females. The assumption under this theory has been that the females are passive, they're not doing anything. It's just like Darwin: the female is passive and coy, and the males are active and responsive.

Morse Δ And there's no female selection going on.

Winthrop Δ Right. That model has been operative for almost 30 years now, and it wasn't until I came along that it was really questioned whether or not the female responds. There are one or two people, and I think this is also significant, the two people who had questioned that model, one of them was from India and the other guy was Asian. So they're not Americans, they're not even Europeans, and they're both men.

Morse Δ Have I heard about your research before?

Winthrop Δ Probably not, because I can't get it published, but that's another story. So anyway, what I was looking at was whether the females responded to female-produced chemicals. And they are. I had 5 years of field data with P values 0.0001 that say, yeah, the females respond when you put out female chemicals. And I had two species in which I showed this.

While I was doing my thesis, I also wrote two papers about the graduate experience for women and minorities in my field. I think those papers got me blackballed, but I can't prove it. I don't know if that's why I really can't get anything published, but I had two papers before that, and all of those papers were accepted with minor revision or just flat-out accepted, and now I keep getting rejects.

Morse Δ Is it a really small community that's doing peer review?

Winthrop Δ Yes. I've gotten one "accept with minor" and one "reject," and then they'll send it to a third reviewer who rejects it. I don't know, it may be that suddenly my writing skills have failed. When I wrote those two papers about graduate women, they were really toned down: they went through ten peer reviews. They just reviewed, and reviewed, and reviewed. They made me take out really stupid stuff. I had in there the percentage of women in my field in 1972, which was 4.8%. And then I had in there the percentage of women in the field in 1991, which was 5%. And they would not let me put that in there. They made me take it out. Don't ask me—I don't know why. Eventually I snuck them [the statistics] into a part of the paper where they didn't attract much notice, but I had wanted to open the paper saying, "Hey, look, it's been 20 years, there's been no change, why is that?" But I wasn't allowed to do that.

I did a survey. I managed to get a list of all the graduate students that belonged to [our major professional society]. I used this mailing list to mail out a survey. I asked them things like, how often did they talk to their advisor, and what they thought of the climate in general in the department, and how often did they see racist or sexist stuff happening in their department. I was surprised [about the results], I guess, but when I really thought about it, I wasn't. It was pretty shocking to a lot of people, especially since I think as many as 10% of the people said that they still heard people using words like "nigger" in their department.

Morse Δ No way.

Winthrop Δ Way. I'm serious. In my department as a graduate student, we had people who routinely used that word to refer to black people.

Morse Δ Well, so, how were women treated?

Winthrop Δ Crappy. But that's because there were no women faculty. There are none. They are making progress; it's coming along, but it's slow. There's also a backlash, because any time a woman gets hired now, men are saying, "It's not because she's really competent. It's because she's a woman, and affirmative action made them hire her." I went through that. I won an award from a society for a paper I presented, and someone came up to me and said, "Well, they just gave that to you because they have to give an award to a woman." People are so obviously competent and well qualified, but men are still saying that. There's one woman in the department here, I was in the committee search and she was far and away the most competent person. She had outstanding qualifications, she's done a great job, and people still are griping that the only reason they hired her was because she's a woman.

Morse Δ Is that going to get worse? Are women becoming the scapegoats in a terrible labor market for scientists?

Winthrop Δ I don't know. There was a conference of an association of mathematics, specifically titled, "Are women getting all of the jobs?" and after having this huge argument they published a paper on the proceedings, where they looked at the statistics and found that women were actually getting fewer jobs than men. But the perception was the opposite, and there was a lot of anger that women were getting all of the jobs. Essentially nobody is getting any jobs, but the women are getting blamed for that, not the job market.

Morse Δ So how long to you think the backlash in going to last?

Winthrop Δ Until I die. Until there's some new group to blame. I don't expect to see it over with anytime soon.

Morse Δ Here's a make-believe question. You can change anything about science. What would you change?

Winthrop Δ Oh, the tenure system. The tenure system is impossible unless you have a wife. Unless you have someone to take care of your home life, and I mean a wife in the traditional 1950s wife sense, the little woman at home, because it's just impossible. Especially for me, because I have a position where I have a big teaching load, but I'm still expected to do research to get tenure. So you get put in this bind, where first you're working at least 40 hours a week teaching and doing teaching-related duties. If you have an exam, then that goes to 50 or 60 hours a week. On top of that, you're trying to fit in your research, which, depending upon what you're doing, can take several more hours. I routinely work 70 to 80 hours a week. I love some of it, but at the same time, I would really like to have a life. And you cannot—you cannot have any other life besides your research and your teaching. It's just impossible. The way the system is set up, I don't think anybody that gets through it is really a healthy person—male or female. I don't think you can be a good, fully functioning human and make it through the tenure system. I think that's why there are so many assholes with tenure!

Morse Δ Do you think that great scientists care if their colleagues are kind and healthy?

Winthrop Δ Well, the assumption is, if you are good enough, then you will make it through. And there are some people who can really live for their science, and live for their teaching, and that's all they really want. I can't do that. I need time off to do nothing or to do things that I'm interested in. When I was in graduate school, I had a time when I did a lot of charity work. I can't do that now, because I'm too exhausted. Teaching is, if you do it right; anyway, it's just exhausting. And research, since I have a lot of different things going on, it's basically just like being forced to juggle 20 balls in the air, nonstop, for 7 years. And some people make it through, but I can't think of anybody I know who is not warped in some personal way who has tenure right now. Even the women I know who are on the tenure track and are succeeding have done so at the cost of their personal lives; their marriages have broken up, or they have essentially a nonfunctional relationship, or they never see their children, and they have no idea what their kids are doing. It's a system that was drawn up decades ago, and it was a system for white, privileged men who had wives who took care of the kids, who did all of the things that kept their lives going, so that they could devote themselves to the God of science. You were supposed to suffer. I think that was part of the proving process, that you could suffer and persevere. In this decade, a lot of people just don't have patience with that. I don't know, maybe it's just 1970s "me-decade" leftovers. But unless I get a really different tenure-track job, I'm really leery of accepting any other tenure-track position.

Morse Δ Would you ever leave science?

Winthrop Δ Possibly.

Morse Δ What would you do?

Winthrop Δ Writing. That would be the other thing I would choose to do. Right now, I'm essentially following an obsession, because I've started this really fascinating project with these organisms, and I'm getting these wonderful results that nobody else in the world has, and I want to know what's going on in this system. I think, and some other people as well, some of the bigger names in the field, think that this could be something really big, a new factor that we need to consider. What I'm really interested in is resistance. If you're going to try to use these chemicals for mating disruption, is it possible that behaviorally the females could change and you could end up with behavioral resistance? It's the same thing as pesticides. It's chemical input into a natural system; it will evolve to deal with it.

Morse Δ You're fairly close to what I consider ethical issues—developing environmental chemicals and pesticides.

Winthrop Δ One thing I won't do is take a job with a chemical company, because that's too much of a sellout. I have a lot of problems with delivery systems, like with a genetically engineered crop. There are seeds that have certain genes in them, so you have to buy the seeds from one company. Then you have to buy a promoter that turns on the gene in the seed, so that's a second spray. And then it's resistant to certain herbicides, so you have to buy the same herbicides from the company. It's essentially sort of like making the pesticide company a company store: You're completely dependent on it for all these chemicals. That's why I got into agriculture in the first place, because I wanted to do something to lower inputs, and I figured the best way to do that was to be the person helping make those decisions about what to spray.

Morse Δ So do you think that scientists have more control over the outcomes of the use of the technologies they create than some people do?

Winthrop Δ I wouldn't say that. I think that my view of nature is that we have no idea what the hell we're doing. We have no idea how it works. Anything we do is going to have a reaction. We may just not see it right now; but it's coming back. I don't want to be involved in developing chemical inputs, but that's just my own personal choice. They are necessary; I mean, there are some cases where you just have to do it. It's the misuse and overuse that needs to be cut back on. But there's a lot of stuff we're spraying that we have no idea what the heck it does. We have no idea.

Morse Δ Tell me about the time you were hassled by your department because of your clothing.

Winthrop Δ When they hired me, they hired five other male faculty members at the same time. One of these guys is just a total slob, he even has food on his clothes. There's another guy who rides his bike to school and shows up in his T-shirt and shorts for class. I never thought very much about what my professors were wearing when I was taking classes, but I wear a skirt, nearly all days. It's just a plain print skirt. I kind of think, well, it's a skirt, I've done my deed. I'm not a real spiffy dresser, but I'm neat. And I started getting all of these comments on my teaching evaluations about the way I was dressing. And it was suggested to me by our dean that I start dressing more neatly. But he never said anything to any of the male professors who were dressing less cleanly than I am, or to all the other people who had been there for years who were wandering around in jeans.

Morse Δ You were held to a much higher standard.

Winthrop Δ Oh yeah. That's been my whole experience here, and it's been awful. I teach a class in evolution, and I've been getting death threats every semester that I've taught it. I'm pretty sure they're from students. They cut out a picture of me in the newspaper and stuck it on my door with horns drawn on my head. I took it down and they did another one and put it up again the next day.

Morse Δ Do your students take you seriously?

Winthrop Δ No. They do now, after I've been there for a couple of years, but I've had to really fight for it. It's been really difficult. I've been forced to be an authoritative figure, because otherwise they wouldn't listen to me. They all look at me and assume that I know nothing. The first semester I was there, I was stalked by a student. First he was stalking one of my students, and I called him into my office and I said, "Leave her alone, this is inappropriate, I don't want you disturbing the female students in my class, OK? So stay away. If you do that, then I won't file any charges against you." So what he did was, he stopped stalking the student in my class and he started stalking me! Then, I sent him to the counseling center, and they talked to him for 15 minutes, called the mental hospital, and had him picked up!

But the university just doesn't give a shit. This last semester, we had one student who was stalking three different female students. He would be in their class and would follow them around in class: He would come and sit next to them in class and they'd get up and move, and he'd follow them and go and sit next to them again, and just like chase them all over the classroom. And then he'd also follow them around campus. But the university didn't have any formal response to this situation. . . . They did nothing.

Morse Δ So even though you did file a complaint, they just had no means of dealing with it . . .

Winthrop Δ Yeah, they had no policy at all. They're *debating* policy, they're *wondering* what to do about it now; but in the meantime, this guy was allowed to stay in the classes with the students, and they finally ended up dropping the class, and one of them ended up withdrawing from the university. He just chased her away and our university did nothing. Nothing.

Morse Δ These were undergrads . . . do you think their future in science is a strong one?

Winthrop Δ No. But I think there are so many cards stacked against these women there, that most of them, the highest they're aiming for is a degree in education. That's as far as they look. Even that is a major step for them. So many of our students are returning women who are older. Almost every single one of them gets divorced within a year of starting school because their husbands just can't stand it. Every single one of them gets divorced—I don't know any one of them that has gone back to school and has stayed together in a healthy marriage. They're in their late 30s and early 40s, most of them have children, and they're really courageous women. We have this mix, it's like 50/50 older students and regular-age students. It's just incredible. Almost all of them have lots of kids, and they're trying to juggle keeping all these kids, and fetching kids, and running them around, and also doing classes. And they're doing very well. I have one student who I'm really trying to encourage to go to graduate school. I don't know if she's going to do it, because she's 45. But she's sharp, really sharp.

Morse Δ Are you ever going to start a family?

Winthrop Δ Well, I doubt it. First it would be nice if I had a spouse, but that's also out. I found that the first really promising relationship I had broke up because he couldn't deal with the fact that I had a PhD and he didn't. And that's been a stumbling block frequently. But I think that's kind of typical of the whole way that men deal with women. They're not used to dealing with them professionally, so they revert to the ways that they do know how to deal with them, which is through the way they look, the way that they dress.

Morse Δ Will that change with the younger generation?

Winthrop Δ No, I think it will get worse, actually. Most of the younger women I see coming up now have really bought in to the "feminism as evil" myth. It's almost like a generation of women who say, "Well, I'm not a feminist, but, I think I should get paid the same as a guy," and they go and espouse all these feminist principles, but they assure themselves that they're not feminists. Most of the younger women I see wear a lot of makeup and worry a lot about their looks. Not that I don't, but I guess once I turned 30, I just said forget it. I think I'm probably just doomed to be celibate, because there are no guys. All of the nice guys are already married and the marginal guys can't deal with a smart woman. And actually, most of the nice guys who are married can't either. I fully expect to go to my grave unwed, but with lots of brilliant research done.

Morse Δ Are you happy about that?

Winthrop Δ No, I'm not happy about that at all, but it seems inevitable at this point. I've started down a path and it's too late to go back.

Martha Gunderson has recently left her engineering position in industry for a job in academia. She is 27, has her master's degree, and works at a large northwestern university in an administrative position, supporting women in science program. She is married with no children.

Morse △ Did you find that when you were working in industry that you were caught in any ethical dilemmas as far as how you did your day-to-day work?

Gunderson △ Actually, the position I held in industry was as a quality engineer, a process quality engineer. That wasn't really what I was trained for, but at the time that I took the job, there were a couple of things going on—this was a very good company and I wanted to get into the company. Also, my husband and I were looking for positions at the same time, and it was difficult for us to find the exact perfect position in industry in the same city. It wasn't exactly what I wanted to do, and I saw it more as an entry-level position; but as it turned out, I think it suited me, because if you think about it, the best quality engineer a company can have is one that is very ethical, and I found myself to be a very ethical person. If this was right, or if it just wasn't the right thing to do, then that was the stand I took, and my position allowed me to take that stand. So it actually worked well with my sort of personal beliefs. Because manufacturing engineers, I'm not sure how much you've talked to people about specific engineering fields, but manufacturing engineers are told by the corporation that you have to produce product. That's your primary thing—you produce products. And so that's what they're concerned with. Sometimes the whole quality aspect of whether the product is going to work goes out the door. And in the industry that I was in, which is the medical device industry, that can mean a person's life. And so the quality engineer has to sort of be the reins on the stampeding horse, and say, "Look, we've got to make good product. We've got to make product, right, but we have to make *good* product."

Morse △ Are there any pressures to relax standards just to get the product out there?

Gunderson △ Oh, there are always pressures to relax standards. And the question of, well, that *might* happen, but *will* it happen? You can come and say, "If we release this product, this might happen." Yes, but *will* it happen? You know, that's always the question. And there's always the push to really be sure: "Are you really sure that this is a problem?" So there's a lot of questioning. The company in particular that I worked for is very good about being concerned about the patient in the long run. So they wanted to make sure that if they had to throw product away, that they *had* to throw it away. But they were willing to do so if there was a patient risk. Now in a lot of other industries, where there really isn't as much of a risk factor, I think that that isn't the thing that happens most often.

Morse Δ Here's a big generalization. Would women have higher standards if they controlled the process, all the way from CEO to consumer? Would they be more stringent?

Gunderson Δ You know, that's really hard to say because it's so individual. I know that if I were personally involved, then I would be much more stringent than companies generally tend to be. I mean, the arguments that I had with the manufacturing engineers wouldn't have happened.

Morse Δ Were they men?

Gunderson Δ Yes, but we're looking at a field where everybody I worked with were men. As it turned out, the interesting thing was, my manager, the manager of quality engineering, was a woman and so was our director. So it seemed like the women against the men. But I'm not sure if that was really the case. You know, it's really hard to say. I could say yes to you, but I'm not sure that would be very accurate.

Morse Δ Any hunch whether women as a group are going into the quality field? Is this going to end up being the "women's" part of engineering?

Gunderson Δ The really sad part about quality engineers is that they're really looked down on in the industry, and I hope it doesn't become a women's field until the attitude about that particular type of engineer has improved. In engineering there's kind of a hierarchy, and it really gets down to a sort of snootiness: you know, certain engineers think they're better than other engineers, and that goes as far as university majors. If you have a mechanical engineer and a civil engineer in the same room, the mechanical engineer is going to be completely convinced that he or she is the better engineer, because that's just the way things are.

Morse Δ A pecking order.

Gunderson Δ Exactly. In industry, it's kind of the same thing. The design engineer is, of course, much better than the manufacturing engineer, who is much better than the quality engineer. That's just how people think of it in general, but I don't think it's actually the case. I thought I was a very good quality engineer. But it's really interesting how that sort of play comes into it, and that sometimes more women end up in those fields that are maybe not considered to be as important on the scale.

Morse Δ Why does that happen?

Gunderson Δ A lot of reasons, I think, that the hierarchy exists is because some of the engineering fields are considered a little bit softer than the others, and that quality engineers don't sit there at their desk with a calculator, and their computer, and do equations all day and design product. They don't do that. They may do some statistical analysis, but for the most part they are interested in sort of bigger picture, global kinds of things and the rules and regulations that go along with building product. Those kinds of things are a little bit softer. You can look at it on an even bigger picture, where the physical sciences and engineering look down on the social sciences. And it's just that on a smaller scale.

Morse Δ So at some point, maybe these women were guided into it because it was softer—kind of an entry-level thing.

Gunderson Δ Or that they ended up feeling more comfortable in it. Personally, I'm not sure that the softer sciences are less good. And that maybe women in general, because of the way they think and because of their personalities get along better with those kinds of sciences, because they have more people aspects to them, and it's just the question of how they're looked on by the men and everybody else in the field.

Morse Δ Is there any old-girl's club that's emerging?

Gunderson Δ I don't see it. I haven't run into it yet. I think that women tend to encourage other women, but I don't think they do that at the exclusion of men. I think women are in general more nurturing, and when they see someone that they think is going to be successful, they try to help them, and that's regardless of their gender, at least the people that I've run into and how I personally act, that's how it's been.

Morse Δ Have you ever had any feelings of being back-stabbed? I have been told that older women sometimes have ridiculously high standards for younger women scientists, because they had to go through so much themselves just to make it. The younger women are expected to run the gauntlet.

Gunderson Δ In the position I have now, I hear that a lot. I personally have not experienced it. I talk to a lot of the younger women faculty and a lot of women graduate students, and they perceive that as going on, but in the field that I've been in, there really aren't any women. In engineering mechanics I don't know of any. It's a small field, and I don't know of any women who have been around for more than, say, 10 years, in the field. But I kind of know what you're getting at, though.

Morse Δ It's the Margaret Thatcher thing. Iron women only allowed within.

Gunderson Δ It's very difficult to generalize, because you're dealing with such a wide range of people. There are some women [who] succeeded in the field because they are probably just exactly like the men. And I think that there are some women who have been around for quite a while, that the reason they succeeded in the field is because they've been socialized to be like the men. I mean, the only way that they could see to succeed was to become the thing that was already there. And I think just finally now we're getting to the point where women can go into science and engineering and still be themselves. They have a lot of battles in the field, but they can still be themselves.

Morse Δ Will men be able to open up a little bit when there are more women in science?

Gunderson Δ We tend to generalize men as a different and supposedly evil thing, but I don't think that's true. I've met a lot of men who are very supportive of the concepts that women are trying to bring into the field right now, because they feel more comfortable with them too. There's also a sort of range of personal behavior or styles in men, and some of those people feel more comfortable with the kinds of ideas that feminist women are trying to bring into science. There are also men who are very uncomfortable with those things.

Morse Δ Do you think that has anything to do with age? Or again just a range of personalities and styles?

Gunderson Δ Well, I think it's a range of personalities and styles, and also sort of the socialization inside the field. In order [for young men] to be successful scientists, they had to become just like the generation of women that we're talking about. They had to become a certain type in order to succeed. I think that there are probably some men in the older generation that would have been much more comfortable being something a little bit different.

Morse Δ What do you think science and technology will look like when and if women make up half of the fields' practitioners and funders?

Gunderson Δ I think that it will probably happen. It's starting to move in that direction now. It will probably be a more collaborative type of field, that people will work together toward common goals instead of battling against each other and thinking that their theories are always in opposition when actually they mesh together. I think that the reward system will also change, and it won't be based so much on individuals and the empires that they build, but rather on collaborations and successful work that you've achieved with a group of people. I think that working together will be more highly respected and encouraged through the reward system. Well, I hope. That's like the dream.

Morse Δ Is there any reason for that to happen if science and engineering cultures don't change?

Gunderson Δ I think really that the driving force in pretty much anything is industry. In this day and age, industry sort of makes things change. I think that industries are starting to see that pitting their employees against each other all the time is physically bad for them. They get ulcers. Industry sees that they will have more productive people if [their employees] are happy in their jobs. And part of being happy in a job is thinking that work is a good place to go, and a fun place to go, and that you're working toward a common goal with other people. In the past, it's been very adversarial, and you're battling against everybody else in order to attain the promotion that you want. It's just very confrontational. I think that industries are starting to recognize this. And when industries start to recognize this, then they start to look for students coming out of school, coming out of the universities, that can work together, that can work in teams, that can work in an environment where people cooperate.

I've had a couple of companies contact me in the position I'm in now. Boeing, for instance, is very interested in making changes at the universities, so that students come out prepared to work in teams and prepared to work together on things. And the universities right now aren't training students to do that. It's almost as bad as medical school, going to school in engineering, medical school is even worse, I think, in the way they pit their students against each other, but going to school in engineering is almost as bad. I mean, you very seldom work in teams, and it's all based on you personally, what you do on a test, and the test is graded on a curve. So that means you're battling against everybody else in the class in order to get the grade that you need. So, I think that industry is going to force the education of their engineers to change. And in that respect the engineers that end up going to industry and the engineers that end up going into academia are going to be trained a little differently, and they're going to impose that training on the next world that they come to. I think that, as industry is changing, in a slower manner, academia will change also. They're going to have to teach differently, and the people that are going to end up going into academia years down the road are going to be trained to work together and to expect to work together on things. Which would be a big change. Now it's unfortunate, because the industries, the ones I've talked to, have been wanting to do this for a couple of years, and they really want to change the universities, but the universities are very slow to change. So it's probably going to take an incredibly long time to happen.

Morse Δ Anything else?

Gunderson Δ It's tough to say whether science is being feminized or if maybe it's just changing. And that it's changing in a way that women in general feel more comfortable with. It's hard to say who's impacting who, and that's just as an aside point. I wonder if science is getting feminized. I'm not convinced that it is.

At 38, Barbara Goshko certainly followed a nontraditional path to becoming a scientist. She surmounted a nearly disastrous high school experience to become a successful theoretical physicist. Dr. Goshko, the mother of two children, is looking for her first permanent position, and was waiting to hear about the results of an interview at a prominent West Coast university when we spoke.

Morse Δ Why did you become a scientist?

Goshko Δ Because of the NSF reports that said we needed scientists. I was looking for a job. I would have chosen something else had I known [about the disastrous job market]. I did decide that I wanted to be a scientist when I was really young, before I even knew quite what it was. I was very good at arithmetic. I was seriously discouraged, I mean, people laughed at me when I said I wanted to become a scientist. In third grade I tested out at college level in math. This was because I played with the stuff; it was a toy. They told my folks that it was an error.

I'm a high school dropout. I had to go to work. [When I dropped out] they gave me fail marks in every one of my classes, and then averaged them into my grade-point average. Most colleges refused to consider me. There are other young people who have real reasons to leave school; I don't have a lot of sympathy for, "Gee, I'm bored," but [these students] are leaving school either because they fear violence, which we hear is happening now, or because they have to go to work. I think they should be given a chance when they want to go on. In my first quarter at college, I had a 3.5 GPA right away. As far as I was concerned, working and going to college didn't seem any harder than working and going to high school. I thought it was funny—they kept telling me how hard college was.

I started out as an accounting major. I took some math and did really well, and was once again very happy. But I had failed physics in high school, partly because I didn't have the proper math background. The rest of it was because I was tired. I was working over 40 hours a week. But I absolutely love physics. I know people who got their degrees in physics, who I sat right next to, who didn't. And I will *never* understand that. If they don't love it, they should go and do something else. I love physics, I'm good at it. I love teaching it, I love doing research, I love reading about it. I've got my family and I've got physics. And, you know, the other stuff that's connected. I don't need anything else.

Morse Δ So, are you one-dimensional? Have you ever been involved in your community?

Goshko Δ I am not involved in the community. I find that about the time they ask me what I do for a living, everybody shuts down. They don't want to talk anymore. Really, seriously, there's a lot of antagonism.

Morse Δ Because they're afraid of you?

Goshko Δ Afraid of physics, is what I would say. There's an image problem with physicists!

Morse Δ Is your life, as a person who loves and is totally immersed in your field, as complete as you want it to be with your family, your physics, and your religious community?

Goshko Δ Absolutely. I am immensely happy. As long as I am employed, I don't need anything else. I love music, I play piano, I go to concerts. I love to take walks. I don't like being with groups of people, that's all.

Morse Δ Is your career going the way you want it to go?

Goshko Δ I don't know what I'm going to be doing permanently. I want to do physics. But like I said, that could be many different things. I want to teach physics, or do research, or applied physics in industry. I am very concerned with what I'll be doing in a year, when my current appointment runs out.

I've received a couple of job interviews. If you could see what happens during those interviews, you'd see how bad it really is. First of all, they have trouble finding a woman to come in to be part of the interviews, so that you see a woman somewhere down the line. The departments, and industry for that matter, go to great lengths to find a woman. They really do. I keep getting asked, "Will I fit in, will I fit in, will I fit in?" This is where my husband and I compare notes. I fit in as well as anybody else, as far as I can tell, in terms of work. Am I a team player when we're doing a project? Of course I am. I do my part, I do more than my part most of the time. But they somehow don't believe that I can be part of the team; they can't picture it.

Morse Δ Do they ever ask illegal questions?

Goshko Δ Yes. All the time. "What year were you born?" was one of the first questions at the interview here. It was asked by a woman! And then she continued on to the relationship between my husband and myself.

Morse Δ What kinds of things did she ask?

Goshko Δ It was totally inappropriate. She accused me of never holding a paying job. That's not true. She accused me of following my husband around. That's not true. I couldn't really tell her, "Well gee, I'd applied all over the country, even though my husband's part of the faculty here, and he'll be following *me* if I get a job." I can't tell *that* to these people. And then she asked the "How can you deal with the children" routine. I saw two male faculty members here, who, by the way, did not ask me anything illegal. I also met with one of the deans. And then, this woman. That's the second time I've been interviewed by a woman in science or engineering, and they both did the same thing to me. I couldn't believe it. It was really bad.

Morse Δ Yes, it is illegal and it is fairly disgusting, but do you think women are unwittingly forming a kinship with you?

Goshko Δ No. I make it very clear that I am uncomfortable. All I can do in my head is think, "How can I get out of here? What can I not say?" People's big thing was probing into the personal aspects of the relationship between my husband and myself. The interview here, where I have a temporary teaching position, I almost walked out on. I actually repeated some of the questions back to her, to make sure. "Is this what you're asking for?" "Yes."

Morse Δ How did you deal with that? Did you answer?

Goshko Δ I answered. You're looking for work, what are you going to do? You're looking for work, the market is difficult, and I did what I felt I had to do to get the job. And I feel humiliated. I don't ever want to see her again. I have interviewed people and it doesn't even *occur* to me to ask those questions. It's the furthest thing in my mind. I want to know, "What's your background in physics, are you interested in this?" I need to know what they know, and try to figure out what the start up time is, and will they be willing to work hard and put in the time that's necessary. I always thought that was the issue.

Morse Δ What is your position on affirmative action?

Goshko Δ I wish it weren't necessary. Everything I see, including the [electronic bulletin board] posts that say, "Put the applicants' information in a computer [and let it determine the best candidate]" won't work. In physics, where women are so underrepresented, there is no other way. If nothing else, it might scare enough people to say, "Well, OK, yeah, we'll let her be here." That's about it.

Morse Δ What's with the club? Why don't they want to let women in? Why is this even necessary?

Goshko Δ I wish it weren't. I think it's just sort of like, well, everything was comfortable. I've been the only woman in many settings. You know, I come in, I am an outsider, I am not comfortable. I came in and sort of wrecked science, is what I get out of that. I didn't fit the mold, they didn't know how to deal with me, and all of a sudden some of the jokes didn't work, a lot of the jokes didn't work. It's no longer a boy's group. A guy's group. A men's group. Obviously, since I'm there, if you allow me or any other woman to stay there, it changes the dynamics of the group immediately. And it's not comfortable. It's sort of like the unknown. They're not used to working with women. *I'm* not used to working with women, you have to realize that.

I think affirmative action is only necessary to a point, OK? To basically give a group the opportunity to get in there, if they're being blocked out. That doesn't necessarily just hold for women in physics, in my mind. That holds for men in certain fields, in nursing and that kind of thing. It's not just for whites.

It's where there's truly an underrepresentation compared to the population at large if you look at the United States.

Morse Δ How do you put in the time that's necessary and still have a good relationship with your family and raise your child?

Goshko Δ I have an office at home.

Morse Δ And the kind of science you do [theoretical physics] makes it possible for you to work at home.

Goshko Δ I'm lucky that way. I don't know how the experimentalists do it. I have no idea.

Morse Δ Do you think there should be any adjustments made for scientists who are stuck with extreme hours?

Goshko Δ Somehow, we've decided that having children and caring about your kids are incompatible with science. Emotion is incompatible with science. We wouldn't allow emotion to affect us. It's putting on airs. And what it comes down to be is an antifamily atmosphere. I've heard from both fathers and mothers on this. They want some understanding on this. The parents want to be able to take a few days off [for a newborn]. A previous employer of mine [a government lab] was wonderful. They were great. I worked from my home. I had somebody taking care of my son in our home too, because you can't concentrate on science and do things like that. But I'd go out [of my home office], nurse the baby, and eat, because I ate like mad when I was nursing, then went back in and continued to work. And when he really needed me I was there. Some of the on-site day care available at corporations is excellent. And not requiring that people be there 7 days a week—my goodness, you're not married to your job!

Morse Δ But I'm being told by some scientists that they are, and most happily, too.

Goshko Δ I'm not married to my job. I feel completely fulfilled, but I didn't marry my physics. I will say that my physics is part of me. That's fair. And that sounds really trite. But I didn't marry it. There's an ordering in my mind on what's important.

Morse Δ Is this different between men and women?

Goshko Δ The men that I get along with, even those that are antiaffirmative action, they also feel strongly about their families. [Some of my male friends] seem to be sacrificing sleep before they'll sacrifice their families. And that's not good. The studies show that if you don't get enough sleep, you're not very creative and you're not accurate.

Morse Δ So what should change for it all to be in balance?

Goshko Δ I worked in Germany for a couple of months. People came to work at 9 and went home at 3. They had a separation between their family and their work. Their work was not deemed to identify them: it was part of what they did. It was part of their life, but not everything.

It's an attitude problem. It seems to be that we don't value our families. I know that there are so many politicians talking about that, and I won't vote for most of them, because I don't think they believe it. Your life is not equal to your work.

Morse Δ How would you judge the science being done in Germany? Is it second-rate?

Goshko Δ No way. Comparing it to my experience at [the top-rated government lab] I would say that overall, the work is comparable. People are less harried. People actually take a real vacation—for weeks and weeks!

I myself struggled with [overwork] a lot, before I was married in particular. I'd stay up at night—I worked, and worked, and worked! I saw my ability to do my work—after I'd had a break—I could see that I wasn't working decently anymore. We also have to be able to say, "No, you're going to lose your creativity!" If you're going to tire yourself out, you're going to make yourself sick.

Morse Δ What you're telling me is very anti-American.

Goshko Δ Oh, sorry. I believe in taking walks around the block, and walking over and listening to a violin quartet.

Elizabeth Lamb is a 27-year-old lecturer at one of the top universities in the nation. She completed her undergraduate degree at an Ivy League school, then went abroad to complete her PhD. She describes her background as being "as normal as apple pie," growing up on the East Coast in a stable and supportive family, with grandparents running a farm in the midwest that she will eventually inherit. Since I first spoke with her, Dr. Lamb has entered into a romantic relationship with another talented professional, whose recent job offer across the country has led to some predictable, but nevertheless difficult, conflicts.

Morse Δ You're 27 years old, you're at a prestigious university, and you're also spending time in business. What kinds of experiences have you had because of being a very young woman?

Lamb Δ It's funny, to be honest. Very interesting experiences, in that a lot of people at the company where I work operate under the assumption that I'm a graduate student doing research; and in a lot of cases I just don't correct them. There's no need for them to know my position, there's no need for them to know that I've got a doctorate. It makes them nervous and it makes them weird.

Morse Δ Are these men and women?

Lamb Δ The group I work in has got quite a few women in it. Very well educated, very nice, a mix of personalities. At the university I'm the only woman faculty in my division. On the part of the school, it's been really surprising because I'm so accepted. The fact that I am a woman has come up probably two or three times, where it's been an issue. One time was with a project director in industry. He is difficult and is pushy, and I think he thinks he can push me around a bit. But the other day I basically let it be known, in so many nice words, that he should back off, and if he pushed any further I was probably not going to be that pleasant. He just has a certain agenda, and I get the feeling that he thinks that he can push me to get his agenda done in a way that he doesn't necessarily do with the other senior faculty. I don't know if that has to do with age, if that has to do with being a woman, I don't know what it is. But he's learning.

The only time [being a woman] has become an issue is one time I was told in my teaching reviews that I was harder on women than I was on men, which I found quite interesting. Maybe I am. And then the person, I'm assuming was a woman, went on to say that I'm a really good teacher. But I found that comment quite curious. It could have been because women are typically not called on, and maybe because I was calling on everybody equally, the women felt that they were being called on more.

Morse Δ Out of their experience.

Lamb Δ Yes. And I'm tougher on students that are weak than on the students that are not weak, because I feel that they need to be challenged. I'm never angry, and I'm never mean to them, but I will push. I always try to encourage them and give really positive feedback, but I will push those students that are weaker, mainly because your natural tendency is to spend time with those students who are stronger. At this university a lot of students are strong. The ones I'm worried about are those that aren't.

Morse Δ So you're making a conscious effort to identify those students?

Lamb Δ Yes, and to really try and encourage them through questioning them and trying to get them to think. I think a lot of them end up just sliding by without much faculty interaction.

Morse Δ Tell me about your graduate school experience overseas. You mentioned to me earlier that you had experienced sexual harassment there.

Lamb Δ The university I attended in Europe is about 20 years behind the times, and there are a combination of effects which combined to make my life sometimes less than pleasant academically. I had a wonderful time there, it was probably one of the best times of my life. I'm really glad that I faced a lot of the stuff that I did when I was in there, mainly because it was across the ocean and it wouldn't affect me now, and I learned how to deal with things. The blatant sexual harassment story was when a fellow student of mine began to feel very insecure about my position in the department: I was finishing my PhD, we had been working in the same set of offices for a length of time. When I first came, he was very much the golden boy. He was doing well, he was the one who was *the* first graduate student in the group, he was "Peter's little baby." (Peter was the head of our group.) He was very much in the lead, and he's always been that way, for his whole life. He'd also gone to an all boy's school. He'd been to a mixed college, but had not mixed with any women engineers up until that point, to my knowledge. He then began falling behind in his PhD. I sometimes questioned the academic nature of the work he was doing, and I would blatantly question it, because people would turn around and question mine. It was part of the working group.

Over the course of a series of seminars that we both attended, my work began gaining international attention. I was getting a lot of incredibly positive feedback from a very large number of quarters about my approach to the way certain systems should be designed. As a result, at the end of a presentation where both of us had presented, I would be standing in the room with four or five people around me, asking me questions about my work. After one particular seminar where we'd gone down south, we had an 8-hour train journey back up to the university. We were all sitting around, and we hadn't had dinner, and the three guys I was traveling with ordered beers. They began to drink, and Charles turned mean while drinking, and began to make sexual, blatant sexual innuendoes about myself and people in the faculty at school. This is the golden boy, in front of two other people in the group, making sexual innuendoes about me and the people we were sitting with. I mean, this is stuff that, in America if it happened, [the perpetrator] would be incredibly seriously reprimanded. I mean, this stuff was just totally unacceptable.

Morse Δ Were there any faculty with you?

Lamb Δ No, but one of the people was a senior research associate, who are equivalent to faculty. Nothing was said by him, but later I got an apology about it. I was angry and upset, but I had to survive the train journey. So I kept trying to change the subject, but Charles kept going back to it, and hassling, and hassling, and hassling, and hassling, to the point where I was physically shaking. It was so upsetting. And I tried to joke it off, I tried all those techniques, and I finally just said, "Look, the cigarette smoke is really getting to me." And I just got up and left, which I hate doing, because for me it's just admitting defeat. But I just couldn't be in a room to take this abuse.

Morse Δ Were the other men participating?

Lamb Δ No, they were just kind of laughing halfheartedly. But nobody really came to my defense, which was tough.

Morse Δ It got worse?

Lamb Δ It got just unbearable! I mean, it wasn't even sexual harassment after that, it was just harassment, where he would scream at me from across the room, he would interrupt me in meetings, he wouldn't give me phone messages, he was just, as my grandma said, "Downright ornery." And it made my time in the office incredibly uncomfortable. To the point where other people in the office began to notice it. And people began to mention it, you know, "What in the Hell is going on?"

I think I threatened every aspect of his self-definition. I was a woman, I was American, I was dynamic, I was basically kicking his ass in research terms, but he thought of it as competition, where I just thought of it as both of us getting good research out of it and learning. I'm not a competitive person. Well, I am a competitive person, but I'm not competitive in those situations where it's in everybody's best interest to have everybody do well.

Morse Δ So he felt you were competing with him directly for some reason.

Lamb Δ He felt very threatened by that whole behavior. All of a sudden, his little place as golden boy was threatened.

Morse Δ Could that kind of experience ever happen in the United States at this time?

Lamb Δ No, I don't think it could, to be honest. In my current situation— I'm sure that women have to put up with this in other situations—but my current position precludes me from ever having to put up with that again. I have to put up with a different type of strain. People are affected by different parts of you, your definition when you meet them, and being a woman has a very visual, very definite impact. Eventually they begin to respond to that percentage of you that really bears impact on them. That may be the fact that you're a woman, that may be the fact that you're great in a certain area, it may be your certain personality type. That's what I found in Europe. It just took them a while to react to the intellect rather than at decent-looking legs. This is not bragging or anything, but I'm not an unattractive woman. I'm always being asked out on dates. People look at me and say, "My God, you're an engineer?" I have shoulder-length red hair, and green eyes, and a great figure, and it doesn't sit well with the sort of professional me. As a result, my femininity makes a very strong impression in a lot of cases. I think that people still expect the stereotypical scientific woman who is straight, mousy, shy, not a very good dresser . . . but then again, that's the stereotype of male engineers. Think about the stereotypical engineer—the nerd, with tape around his glasses, a pocket protector—when you translate that to the woman equivalent, then that's what they expect.

It's interesting, when I hand out my business card, the fact that I am from this prestigious university bears so much weight in so many people's eyes, justifiably or not. It seems to completely override, outstrip, yet not erase—but it's such a dominant issue in any initial contact, the fact that I'm a woman is minimal compared to that.

Because there are no women faculty members, there are very few of us, I think the women students would do better to have a lot more mentors. But there is a very strong women's graduate group here. They do a lot of events with undergraduates. They're a very strong society unto themselves. And in talking with them, the issue came up recently, where the treatment of women at this institution became an issue. So we had to find out how women were treated on very short notice. The overall impression across e-mail was that women were treated very well, there were some issues and instances, but you get those anywhere. But overall, it was a very supportive, very encouraging environment.

Morse Δ What's interesting to me now is the level of angst that women are expressing during my interviews with them. They believe that they have pretty much signed their lives away; because they see that for a successful scientist, child rearing and an outside life between the ages of 22 and 35 are pretty much impossible.

Lamb Δ (Laughs) I have no private life. I have no social life, I mean very little. If you want me to go on about that, oh, yeah, I agree with that. I joke with my students. They didn't understand that when I complain about the fact that my apartment is a mess, and I have no milk, or when I wake up in the morning and my milk is crawling across my refrigerator because I haven't been shopping in ages, you know, [that it is because] I work 80 hour weeks. You know what I need? I need a wife! Now that is going to become an issue at some point. I talk with a lot of senior faculty, about what their wives do, and most of them are married to very educated, very bright, bright women, who stopped working when the kids were born. More than once, somebody has said to me, "Well, I couldn't do so well in this job, and have this job, if my wife worked, because there just isn't enough time in the day, and one person needs to be at home. As a result, we made a conscious decision to do it in this way." And then they look at me, and sort of grin, and shrug their shoulders, like "Have fun figuring that one out." It's tough. It's brutal. However, this university's got a great policy. If you want to turn the tenure clock off for a year if you have a child, you're allowed to; it's got a very good maternity-leave policy, it's got a very good flex-time policy.

But it's not even that you don't have a life between the ages of 22 and 35, you just don't have a life. Full stop. This university and the work you do dominate every aspect of your life. Your whole life is not safe, your vacation is not safe. It's virtually impossible to go away and sever the ties completely for any length of time. My students and fellow faculty members are horrified that I am not bringing my computer on vacation, and that I will not contact e-mail, and that I will not contact the phone. If there's a problem, my secretary will determine whether or not it's an emergency and at that point will start tracking me down. She's the only person who has the numbers where I'll be besides my mother. They know I'm in Milan and that's it.

It scares the hell out of me, to be honest. What's going to happen in the future if I ever want a family?

Morse Δ Why would anyone choose to do science, if that was the guarantee of the lifestyle?

Lamb Δ Because I love what I do. I absolutely adore my work. I love teaching—the biggest thrill that I get is when I'm teaching a student and they finally get it. I just get such a kick out of working with them. And it's sort of like the surgeon who does surgery for 32 hours. They don't want the money, they do it because they love it. OK, well the hours sort of come along with that, and there is just so much opportunity for teaching and for work, and for interesting research, and for interesting, interesting interactions with people. . . the price you pay for that interest is your social life.

Morse Δ At 27, it is a social life. At 35, are you going to pay that price to have a family life?

Lamb Δ I don't know. Currently, I can pay it. I don't mind too badly. Oh, I complain bitterly about it, but I love my work. I really love my student interaction. It's instant positive feedback. Watching them grow, and improve, and discover, and become better people, or scientists, or researchers, because of my input is so great. It's not even so much of an ego trip, but it can be an ego trip. In a way, it's also really rewarding. It's like having 20 children. You really love helping them learn. I don't have to clothe them and diaper them. I am somewhat involved in their private lives, however. I always know who's having marital troubles, and who's girlfriend just dumped them, and who's having problems with their roommate—that affects their work just as much as anything else, and so I tend to keep track of what's happening.

Morse Δ Do you think men do that?

Lamb Δ No, they don't. There was a recent example, where a student was having problems, and the faculty were really getting on his case, and I said, "Well, his mother died last week, and he had to fly out early and cancel his vacation." And they're like, "Well, how do you know that?" And I said, "Well, because he told me. He didn't tell you guys?" "Well, no." So, being the woman, I tend to get all this stuff, which adds to my work load. I'm definitely not in this job for the pay, and I'm definitely not in it for the social life. I'm in the job because it's fun. It's like a big intellectual amusement park.

Morse Δ What do you think about Penelope Leach's positions on the drawbacks of day care for very young children?

Lamb Δ I agree with her. I read Penelope Leach's work, and I'm also aware of other work that says the child needs to have a consistent set of faces and a consistent set of people, who will give positive feedback, and reinforcement, and affection, to give that child security. Where the child is constantly held, the child is not in a crib; it has a constant interaction with one or more sets of adults who are consistent and predictable to the child. For me, my mother worked, but I had a nanny.

Morse Δ So you were raised by your consistent person.

Lamb Δ I was raised by my consistent person. And my mother always played a very strong role in my life. As I grew older, I always had day care. When I was old enough to go, I told my mom, "I want to go to nursery school, I want to be with other children." I didn't want to be alone with a lone day-care person.

Morse Δ How about a woman who is teaching at a small midwestern college, and is making $30,000 a year, and maybe her partner is making $25,000 a year, and they're not able to afford a nanny, and she doesn't have tenure yet . . . they're depending on a child-care provider for her to have her scientific career at all. What can we do for these women?

Lamb Δ Flex-time. And it's not just for them. It's not just what we have to do for women, it's what we have to do for people. I've got two male friends whose wives are about to have babies, and they're taking paternity leaves. And I think that's wonderful. They're spending the first month at home with their wife and their baby. They have arranged flex-time, they have arranged all these things. I think what has to change is the whole attitude toward families in general. The burden for adjustment has always been on the woman in this society. She's the one who needs to take maternity leave, she's the one who needs to alter her career, she's the one who needs to change things in order to make it work, while the man toodles along quite happily in his 9-to-5 high-pressure business meetings. I think what has to be changed and what is changing, especially with my generation and a lot of the men I know who want to be involved with the raising of the child, is that they're beginning to not only demand equal rights, but taking them. And taking the paternity leave, which in some cases, even though it's available to them, it's not encouraged that they take it. And they're taking it anyway. I'm really encouraged in that aspect. And I plan on finding somebody who is sensitive to those issues. I've got quite a criterion list, I don't know if I'll ever meet someone who can meet those requirements!

I think day care was a good first step. Now we have demanded the right to day care. In every industry and every corporation every process is of continual improvement. You've got where you are, and you've got where you're going. And you can never be happy where you are, you always need to be striving for the next step. We've gotten day care, and in some cases we haven't gotten it and we need to fight for it; and I do agree that we need to have something, I mean, having something's better than having nothing, but I think that we need to challenge those basic assumptions and not just be happy with what we've achieved, but to continue to move on, to demand more rights for our children, because in the end, it's going to be a good investment. Couch it in the terms of continual improvement. It's not that we're bashing day care, but what is the next step?

6

Purse Strings and Politics
Interviews with Women in Science Policy

Our sciences are being harnessed to the making of money and the waging of war. The possibility of alternate understandings of the natural world is irrelevant to a culture driven by those interests.

Feminist philosopher Helen Longino[1]

Scientists, and those who exert control of one kind or another over the production and application of science, have a responsibility to the public and to the earth itself for the outcomes of their work. We've seen in interviews that today's scientists, women and men alike, agree or disagree with this idea to varying extents. In any case, the outcomes of science are quite visibly and intimately bound with our cultural, political, economic, and personal interests. In this chapter, we'll talk with five women involved with science policy, either in influential political roles or as science activists.

Dr. Lynn Goldman is Assistant Administrator for the Office of Prevention, Pesticides and Toxic Substances for the US Environmental Protection

Agency (EPA). She received her bachelor of science degree in conservation of natural resources from the University of Houston in 1976, a master's of public health degree from Johns Hopkins University in 1981, and her medical degree from the University of California. In addition to her work as assistant administrator, Dr. Goldman serves as cochair of the EPA's Science Policy Council. She has had a rich career as a pediatrician and epidemiologist and in environmental health research.

Morse Δ Did you go into medicine and into environmental work thinking that you would be working in policy?

Goldman Δ No, I didn't think that. I think that my original career objective was to be an epidemiologist. I studied at the Johns Hopkins School of Hygiene and Public Health, and I got very interested in epidemiology. And then I went to medical school, and then I did my pediatrics training. Then I immediately got a job as an environmental epidemiologist. So I pictured myself at that time really more as a scientist and a clinician than as a policy person. But I did gravitate toward policy issues. One of the things that I really didn't expect in terms of my work as an environmental epidemiologist was that so much of it would involve working with communities. In California, in my public health work, I was conducting studies and carrying out investigations, [many of which] were prompted by concerns of community members. And they were very concerned with understanding what we were doing, and why we were doing it, and what the conclusions meant for them, and particularly what the conclusions meant in terms of cleaning up the environmental problems in their communities. I think that a lot of what pushed me into the policy area was seeing the connection between the research that I was doing and the feelings that I had about wanting to further the objectives of public health and environmental protection.

Morse Δ Do you think, from knowing your colleagues and having this experience, that your feelings about your career have anything to do with being a woman?

Goldman Δ I actually do. One of the things that I thought was very interesting, actually, was in working with the communities, so many of the leaders were women.

Morse Δ The environmental leaders.

Goldman Δ Yes, yes. And the neighborhood activists. In many cases, [there was] almost a profile of women who were very concerned about their children, very concerned about the public health of the community that they lived in. On their own, without a lot of technical education and training, they'd learned about the environmental problem. They had done research in the local library, they had written to the various agencies for technical reports, they had persisted in getting the knowledge on their own, and had used that to form neighborhood organizations, in order to interact with the government and try to get more action for their communities. As I said, it was almost a pattern of these women becoming leaders.

Morse Δ Did they hold down jobs too, or was this a new kind of volunteer thing—where women used to volunteer for the Junior League and now they're becoming environmentalists?

Goldman Δ In a more affluent community, maybe they were Junior Leaguer types; in a less affluent community there were women who perhaps had jobs or perhaps were housewives. But for many of them it was totally new; they had not been leaders in any sense and had become leaders because of their concerns, and you'd watch them evolve, developing the courage to speak up in a meeting in a public gathering, to truly become leaders. Some of them I know are becoming leaders on the national scene. I think that perhaps I, as a woman, could relate to them in a way that may have been different from some of my male colleagues. I had a lot of feelings of empathy for what they were experiencing, in terms of having to live with uncertainties about environmental exposure and what that meant for them in terms of raising their children in their communities; whereas I saw some of my male colleagues, the way they would handle those uncertainties was, "Well, then why be upset? If you don't have a known hazard, you shouldn't worry about it." I found it easier to empathize.

Another thing I've observed is that women feel, more than men, that they have to prove themselves at one level before they feel they deserve to be elevated to another level. I don't know how to say this exactly, but when I look at my own career, I think perhaps if I had been a man, I would have thought right from the outset that I was destined to be in public policy issues. But I didn't. My sense of destiny was that I was going to do the best job that I could do with the job that I was in. When I really felt grounded in doing that, then I felt confident in assuming more responsibility at a different level or a different kind of responsibility. And I have experienced some kinds of men who come into government at that same level, who immediately articulate, you know, "I think I'm going to be director of this department someday." You know what I'm saying? That would not have been my articulation of where I was headed. I think that generally that's still a difference. I'm not saying it's an innate difference, but I think that women really do feel that they have to prove themselves. There's still a sense that it would seem overly pushy and aggressive to say, "I'm going to be in charge of this someday."

Morse Δ What kind of women are coming into EPA now?

Goldman Δ You know, I feel I'm in the middle in a way. I'm 43. There is a generation ahead of me in environmental science and in medical science who I think had it much tougher than I. Women really had to not only prove themselves to be accepted, but also I think had to give up a lot more in order to gain acceptance. I think that women in my generation were able to hold on to more of our femininity and our identity as women. There were more women around, at least for me, a critical mass of women, that I was able to have a lot of peer support that the generation of women older than me didn't have. For many of them, their only peers were men, and they had to fit themselves into the way men are socialized to behave in the workplace. I think things are even better for younger women. Women just coming into the agency now can clearly see women in positions like mine who identify with other women. I don't have a sense that there's something unique or special about me, as a woman, that I'm in this job. I think that I've always had a sense all my life about my abilities, and I work hard, and I can achieve at a pretty high level. And I feel that that's true for a lot of women, and as many women as it is for men. And I like working with and I like mentoring younger women, as well as younger men. I think that there are not as many of us, we're still not proportionally represented, but there are a lot of women at my level in science who can serve as that kind of role model. Which is really different from the kinds of role model that I had. I would often have trouble thinking, "Who can I really identify with?" I really didn't feel that there were many women that I could look at and say, "That's what I want to be like in 20 years."

Morse Δ To change course a little bit: The child lead-poisoning work you've done and the cancer cluster investigations—they all seem to be very human-centered things, and in some cases, I don't want to say you are advocating for victims, but for the people with the least voice.

Goldman Δ Yes, that's right. And that's one of the other reasons that I went into public health as a career. I've always been very concerned about equity, the circumstances of people in our society who are disadvantaged either because of poverty or discrimination or other reasons, and particularly about children in those situations. When I was in medical school, I was involved with a community group in Oakland called the Coalition to Fight Infant Mortality. Oakland had a very high rate of infant mortality in a very lower-income black community. What we were doing was pulling together the medical and epidemiological information and getting that out to people in the community, to help them put together efforts to get the programs in place to prevent infant mortality. Those kind of issues have always tugged at my heart. You probably also noticed that I was a mathematics major initially in college. And I loved that. I loved—believe it or not—I loved theoretical mathematics. It didn't have the kind of human side that fits me, and I came to realize that epidemiology was a good place for somebody like me, because it's a way to combine that kind of fascination with numbers and teasing apart difficult and complicated problems with a concern for people and their lives and being able to do something to help improve people's lives. I think that's how I ended up where I ended up.

Morse △ Do you ever sense that there's a concerted enemy out there? In the environmental field, I sometimes hear, "Well, it's industry, and the captains of industry, who are fighting any environmental action or efforts to clean up dumps." Do you find any truth in this?

Goldman △ Well, if there is an enemy, it's in all of us, in that all of us, on the one hand, have very high expectations of the environment and, on the other hand, have many things in our own behaviors that contribute to the problems in the environment. I don't think that it's a black-and-white thing of industry versus environmentalists at all. I do think that much of what we have to do for environmental and public health protection is to change people's behavior. And some of that behavior is economic behavior—it's manufacturing, it's production, and so forth, but some of it is also personal behavior, it's whether we drive the car or take public transportation, it's whether or not we recycle. I really think that we all have to work together to develop an understanding of what we want to achieve, what we want this world to look like, not only today but also in the future. I feel very strongly about that issue of our responsibility to future generations. I mean in a way, if you want to say what the enemy is, we are all the enemy, when you think about our grandchildren and our great-grandchildren. Who are they going to be able to point the finger at if we don't do major things to change our behavior? I really think that the whole problem has been miscast as the conflict between industry and environmentalists, because many in the industry want the same thing that I want. I think that everybody is concerned, not only about the environment they live in, but also the environment their children are going to inherit. What is often missing is a common understanding of what the goals are and how we can get there. We need to establish a playing field for getting there that's a fair and level playing field, in terms of the industry.

Morse Δ What do you think government is going to look like from the environmental policy standpoint when women make up half of all political appointees and half of all professional staff in federal agencies? It's a fantasy, because I don't know if it's going to happen in my lifetime, but if it did happen, what would change?

Goldman Δ Boy, I would hope that it could happen in our lifetimes, actually. I think one of the main things that would change is that government could work in a less adversarial fashion, and be more involved with building partnerships, and achieving consensus, and moving forward with consensus, instead of getting where we want to go through adversarial relationships. I think the record for environmental protection is fairly clear that where we've gotten bogged down in adversarial conflicts, we've often not made very much progress. It's been one of the problems with the Superfund process. It's certainly been one of the problems with the Toxic Substances Control Act implementation, that the extent to which we fail to bring people together to achieve a common purpose and to move forward as a society, we don't make progress. I think women are good at doing that. Women are good at forging partnerships, they are good at forging consensus. Women can be almost unstoppable when they've become determined to do that. They don't easily give up!

I also think that women tend to focus more on children. I see a lot more interest among the women who I work with in children's issues. I don't think it's an accident that many of those who have worked in areas like child development, birth defects, and lead poisoning have been women. Also, women's health issues—making sure that we begin to address the risks for women. One of the most important cancers, breast cancer, which will probably affect one in eight of us sometime in our lifetime—we still don't know the cause of breast cancer for something like four-fifths of the cases. We don't know what causes it. I think there's a need to bring more attention to the things that kill women, as well as the things that kill men, which we've been looking at for many years. I'm sure that if testicular cancer affected one in eight men, we would know the cause. We wouldn't be dealing with it by just having men do testicular self-exams and early identification, because men would understand that it's not enough just to identify it early and have a testicle removed. They'd understand that you'd want to do primary prevention, to prevent it from happening in the first place. And I think the workplace is different with more women in it. Women value team building, and certainly my deputy here, who is a woman, places a very high value not only on individual achievements, but on achievements of teams. One of the things I've been learning from her is how you reward people for team building and for the accomplishments of their group, as opposed to just the individual accomplishments. It's really a nice thing. It changes the atmosphere in the workplace, when people are falling all over themselves to cooperate with each other and get the job done, as opposed to competing with each other.

Morse Δ Do you think that could hold true for the sciences? I've been asking researchers whether they think that good science and good research can be done on a team basis.

Goldman Δ In my field, in epidemiology, it is very hard to go off on your own and accomplish anything, because almost every investigation, especially in environmental epidemiology, is a multidisciplinary undertaking. Those who really excel are those who are able to build strong teams, as opposed to those who are sole investigators. I think in other fields it may be very different. I think that in general, in science, and especially outside of government, things may be different. Now, I think a certain amount of competition is very healthy, don't get me wrong. I am not against competition at all, because I think that scientists need a certain amount of pressure to thrive. They certainly need to be put on the spot—I think that some of the best thinking, for me, has been in the context of preparing to present my work to a group of peers that will be very critical. When I'm under that kind of pressure, sometimes I do my very best thinking. I'm not only thinking, "What does Lynn Goldman think," but I'm also thinking, "What does Dr. So-and-So think?" It forces me to look at my data, look at the issues, from somebody else's eyes. At the same time I also think that you can take competition to such an extreme that it becomes unproductive. Increasingly, in the work that we all do, we need to involve people from a number of disciplines. And that just doesn't go over very well if you're in competition with them. A bit of competitiveness provides the creative tension that you need; but on the other hand, you can take it to an extreme.

Democratic Congresswoman Anna Eshoo was first elected to the US House of Representatives in 1992, and was reelected with 61% of the vote in 1994. She represents northern California's 14th Congressional District, located between San Francisco and San Jose. Her district is home to "Silicon Valley," an area that boasts a large number of high-technology companies, as well as Stanford University and the NASA/Ames Research Center. Congresswoman Eshoo serves on the House Committee on Commerce, which has policy-making power over high technology, biotechnology, public health, and the environment. She has been a vocal advocate on Capitol Hill for the advancement of women scientists.

Morse Δ What do you think about the push to get more women into the sciences, given the current glut of PhD-level scientists? Is that something you think about or are formulating any policies on?

Eshoo Δ I think you have to move back to why there are so few [women in science] in the first place. This is not a field of professionalism that is one that girls and young women have been encouraged to move into for generation after generation. It is an area unlike others, where there have been tremendous breakthroughs, such as in law and medicine. This is an area that still has relatively very few women. It goes back to, I think, how girls—not young women—but how girls are taught. There's still much to be done about it. It's an issue that I've spoken to and brought up continually since I took a seat on the Science, Space, and Technology committee. In fact, it was the subject of a hearing that we had. We had expert testimony from all over the country, brought to the table by women. I also raise the point that in all of the experts that come before us at Science, Space, and Technology, where are the women? Staff was not even reaching out to bring women in!

Morse Δ Why not?

Eshoo Δ It's just business as usual. Now, they've been sensitized. I don't think it was a plot. They just didn't think of approaching it in those terms. There was such an absence—it was startling to me when I arrived here—that not only was the room filled with men, both behind the dais and in the audience, but all the witnesses were men. Staff people would sit behind them, and many times the backup people would be women, of course, holding the keys to so much of the knowledge. There are many, many ways to go at this, but I think that as you work with and learn from the women who are already there, and acknowledge how few there are, that it underscores the need to do so much more in how girls are taught. This is indeed a field that should be considered by them. They need to be encouraged: it's how they are taught. Thank goodness we have documentation. The American Association of University Women came out with a national report, a study that had been conducted for a decade on this very issue, so it has a great deal of credibility behind it. Certainly the Congressional Women's Caucus in the House has taken that and run with it. We've been successful with some language, having some feet on it, going over to the Department of Education. I think that we continue to chip away at it.

Now if there are more women who are elected, we can certainly be considered role models. But we need more women in science! I think that in some ways the jury is still out, because we don't have enough women there to say that "this is how it would change." My sense is that going beyond simply the information that is garnered and studied, you know, brought forward from the research that the scientists do, that women have a tendency to broaden out the applications and the betterment of humankind. I think that is already the case for many, but I think it is at the base of how women think and what they want to do with whatever it is they are doing.

Morse Δ So women have a more humane sensibility.

Eshoo Δ Yes, I think that they have more of a global and pragmatic sense of the improvement of the condition of humankind, whatever that area may be. There is that overarching sense. Now, that's not to say that our male counterparts don't, but I think that women are very clear about that. There's an outstanding organization, the Association for Women in Science—they've been a terrific partner to me on so many of these issues, because they have gathered the data and they have the network of women all over the country. For example: When I discovered that the recommendations that the National Science Board makes for the seats on the National Science Foundation, the National Science Foundation being the highest body in our nation, of the 22 people on the board, there was one woman on the entire board, and that term was running out. That is just totally unacceptable. Now, I'm not looking for slots and numbers, but it more than suggests that, at best, we may have to scrape the bottom of the barrel to see if we have any women in science and see if we can get them on [the board]—and that's just not the case. It's an area that's been totally overlooked. Every time that comes up, I'm on it at my committee. And I very purposefully sought, for more than one reason, a seat on Science, Space, and Technology. If you were to put out to the general public, "Do you think that women legislators would have a tendency to be much better on issues such as education, the environment, children, social issues," I think hands down the public would say yes. If you were to bring in science and space issues, I don't think that those numbers would be able to sustain themselves. Why? Because we're not thought of as being part of that field and having any kind of expertise. Now, obviously, the Science, Space, and Technology committee has a great deal to do with my district, and so that was one of the key motivations to secure a seat on the committee, but I also see it as being our future, and that's what I want to keep my eye on; not the past but our future. Also, these are the areas where people think we cannot distinguish ourselves. You see? That's another challenge for me. I'd like to think that I've made a difference on that committee on these very issues and many others as well.

One of the major accomplishments of my first year of my first term, from the get go, was to meet with the baby-new administration. Within days of the President's inauguration, I was over at the Old Executive Office Building with representatives of the White House, of the Vice President's office, and others, to advance the idea that the President's budget should reflect dollars for high-energy physics. Obviously, I had my own agenda on that being my district's agenda, and the Stanford Linear Accelerator and the B factory. But before I could ever advance the latter two things, I thought it had to be a priority in the President's budget. We accomplished that, so when the budget came to the House, it did contain dollars, and then went to work here in the House, and then on the Senate side, and then of course to reinforce it with the administration. We were in direct competition with Cornell University, and with Senator Moynihan, and the rest is history. This is a major, major accomplishment for California. It's a big win for California, it's a big win for Silicon Valley, and it's a huge win for Stanford.

Morse Δ How do you think people react to you, as a woman? Do your peers in the Congress respect you?

Eshoo Δ I think it's up to each one of us to demonstrate that we're serious about our issues, that we're knowledgeable, that we know the issues inside and out, and that we can compete, and compete hard, not only on the merits of the issues but on anything else that may be attached to it. In other words, you have to do very good work in order to earn the regard. I don't think people just give that to you. In many ways, life is a continual test. Now, do I think we have to work harder at it? Absolutely. But I don't really ever mind the standards being set at whatever level, because I think my constituents deserve that, and I always think to myself, "Just show 'em!" But it is an area especially at Science, Space, and Technology, both on the full committee and on the three subcommittees that I'm on, that I first of all need to know what I'm talking about and be willing to do the homework—I mean, you can't just show up! In many cases showing up is an awful lot, because there are always members that just don't. But I learn a great deal from participating in the hearings or just listening. I don't have to be talking all the time. We have excellent witnesses that come before us. It's very rewarding.

Morse Δ I'd like to read you a quote from a retiring congressman. Tim Penny recommended that, all else being equal, one should vote for a woman, because, "Men play games. We come at this job with a sports mentality, scoring points on the opposition. We have a very competitive approach to public policy making. Women are problem-solvers, by and large. Even if they haven't created the mess, they're willing to help clean it up, while men point fingers and place blame."[2]

Eshoo Δ Well, I think that we're consensus builders. I think that comes from the different hats that we've worn and the different roles that we've played. Women have served on so many committees, you know, in their communities. And when you're on a committee, you have to work with other people. It's not just your idea. You may have a very good idea, but someone else may bring up another point, where you need to slice something off, or add something on, or rework, refine, retune—I think that [women are] products of that kind of a process. We have had to deal with building consensus in our own families; with our children, those of us that have them.

Morse Δ Is that an effective way of moving policies forward?

Eshoo Δ Oh, I think that it is, sure. I think, too, that taking off on what Representative Penny said, when you're a consensus builder you are not so likely to put your dukes up, and say, "I win, therefore you lose." You try and create something in that consensus where each side wins. I think there's a stylistic difference. That doesn't mean that we're softer, that we can't be tough. That should not be the connotation. I think sometimes it is.

I think that women distinguish themselves not only in how they work, but in a variety of places and ways. They are not just simply the traditional ways, as noble, and as good, and certainly as important as those traditional ways are for the people of this nation, because so many of these issues have not been tended to, and I think that's why the country's in the shape that it's in now. We've got a lot of deficits; not just the [budget] deficit that we've been trying to chip away at. For myself, I think it's important for us to be highlighted in the work that we do in the financial pages of the newspapers. I'm not attracted to the social or the society pages or to some of the other columns. I think that they're important—certainly education and the environment— God knows I'm a fighter in those areas, and I'm damn proud of my record. I'm just not a go-along person.

Morse Δ How are we going to create jobs for scientists?

Eshoo Δ Well, I think that the role that the federal government plays in research is front and center. Almost 90% of those dollars that went through the universities for research during the Cold War were applied to and toward defense. We know that's diminishing now, and so the opportunities are still there relative to the dollars, but they need to be fought for and retained, and that provides the backbone. Outside of the university system, and the researchers and the women that are there in the private sector, how do we apply their wit and their intelligence that has been used for defense purposes now for civilian purposes or dual use? There are a number of initiatives that we've not only authorized at Science, Space, and Technology, but also that the appropriators have appropriated dollars toward. We have so many initiatives! The private sector, they constantly come through my office and say, "we need more of this." We are not only seeding these new opportunities, but are helping to change how we do business. I think it's all toward a better twenty-first century. I have to tell you that.

Dr. Jane Rissler is Senior Staff Scientist with the biotechnology and agriculture program at the nonprofit advocacy group the Union of Concerned Scientists. At 48, Dr. Rissler is recognized as a national expert on biotechnology policy issues. Her experiences in academia soured her on an academic career, and she has thrived in applying science to public policy. Some of her views were included earlier in Chapter 1.

Morse Δ How did you get involved in the environmental field?

Rissler Δ Let me just tell you the track that I took. I got a PhD degree at Cornell, then did a postdoc for about a year and a half at the Boyce Thompson Institute. After that I had a faculty position at the University of Maryland for 5 years. It was at that point, [and it was] actually an accumulated realization that I didn't want to work in a university. And we can talk more about that, because it had a lot to do with sexism. That wasn't the kind of work I wanted to do. I didn't want to do research; I actually went to a career counselor to try to figure out what I wanted to do, and it turns out that part of my realization was that I wanted to work more where there was an intersection of science and public affairs. What I did to make that transition was to get a fellowship from the American Association for the Advancement of Science for a 10-week fellowship at the Environmental Protection Agency in Science and Public Policy. And while I was there, I was so fortunate to work on this new issue: biotechnology regulatory policy. And it turned out that there was actually an opening at EPA for a science advisor to their new biotechnology project. And I was able to get that job. I spent 4 years at the EPA working on biotechnology regulatory policy. After 4 years, a job became available in the public interest community in biotechnology regulatory policy: it was with the National Wildlife Federation (NWF). And believe me, there were so few people working on this issue that I realized it was a rare opportunity to get into the public interest community, which is something I had wanted to do for some time: to work on, to think about science-related issues from a public interest perspective, and to try to do some good for a vastly underrepresented segment of the population. So about 6 years ago I started working at the NWF, and just last fall we moved, my colleague and I, from NWF to the Union of Concerned Scientists. So it's sort of a combination of my liking science, my wanting to do some more immediate good as far as human beings were concerned, and the unattractiveness and unfriendliness of academia.

Morse Δ Tell me about the unattractiveness, the unfriendliness, and perhaps the sexism in academia.

Rissler Δ There is no "perhaps" about it at all. And it is, of course, subtle. There are a couple reasons for the unattractiveness. I'm not sure I understand them, but one of my first realizations, even as I was in graduate school, was that I didn't particularly enjoy doing lab work. That has very little to do, of course, with sexism or anything else; it's a personality trait. I just didn't enjoy the tinkering that is required. But I did it well and I was successful eventually. As I looked around, and I always thought about an academic career—it really hadn't occurred to me to do anything else at this point, because I thought it was a good place to work with young people, to actually be a successful researcher. So I stayed on the track of getting a postdoc and then going into academia. What I of course realized in retrospect is that all of this was so dominated by white men; so many white men controlling this whole process. And what I haven't really analyzed is the extent to which what I didn't like about research was in fact what males often like to do—the tinkering—so I don't know, this business about how research is done, I don't know how this is affected by the male dominance.

In terms of the actual unfriendliness: I was at the University of Maryland and I got a job in a department that was dominated by older white men. There were two women on the faculty of botany who were on the tenure-track position, and I was one of them. There was certainly a desire to have women on the faculty, but once they had them, they didn't know what to do with them. And it was clear to me in retrospect, and as I watched young white men come in, that the young white men fit in in a way that I could never fit in. And that this translated into kinds of advice and support that I would never get. And didn't get. In fact, it became more hostile than that, because I was in the process of trying very hard to improve the quality of the faculty there and ran up against the old white males in a way that I was never forgiven. But if I even looked above that level, the university was just dominated by men and the way that men do business. So while it's difficult to articulate, I'm sure you've heard many women say it's just not the warm atmosphere that you would find in a women's studies program, for example; the relationships among colleagues, the kinds of interactions that one has. And a lot of this, I didn't even realize at the time, I didn't realize that I was threatening these old white men. Because I was coming in somewhat aggressive, I was just like all the other men were, but they could not deal with it. You realize, of course, that this is all one-sided; I don't know what they would say, but I wasn't thriving. There wasn't the cooperation, the nurturing, and in fact as it turns out that the only thing I miss is the ability to play squash. So you see, one could never document all of these and bring a suit that says there is sex discrimination in this department. It would be very difficult to bring all of this together. But that's certainly what I felt. And certainly I have not felt that kind of atmosphere to the same extent since I left there. The EPA, it was certainly still dominated by men, [but] there were a lot more women and a much more accepting atmosphere. Of course, in the public interest community, which I hasten to add is still dominated by white men—most of the major organizations are run by white men, most of the upper management is composed of white men—there are some inroads at lower-management positions for women.

In fact, it's young men where the real hope is for an end to the sexism. I don't have much hope for middle-aged and older white men. But I have a lot of hope for how things will change when institutions are filled with [younger people].

Morse Δ From your perspective, are cooperation, competition, and intimidation a part of science?

Rissler Δ In my experience at Cornell with my major professor, that was not the case. It was the case that it was a hardworking lab, but it was not competitive within the lab. We didn't have any postdocs in that lab; sometimes that raises the degree of competitiveness considerably, and that was not the case here. Cornell was pretty intense in terms of doing good research. As my main professor said, "At some universities you have to do research, and at Cornell you really have to prove that you can do good research that makes a contribution," so it wasn't just a matter of doing research and getting out. It had to be well done and show something important or fairly important. I didn't have a sense, in that lab, of competitiveness. Let me tell you how I felt. It was the sense, at the University of Maryland, that there was so little support for doing research. My first year, I got some lab equipment and $600. One had to compete, within and without the University for money, grants and that sort of thing. [You had to] compete for money to do research or you just couldn't be competitive at all. It was much more subtle than having a vicious postdoc stabbing me in the back or something like that. It's the difficulty of establishing a laboratory, and I want to say that there was no sex discrimination about what beginning professors got to start their research; it was just generally bad. What is hard for me to measure, or to even know about, is how much sexism is involved in granting in the major national granting programs, or how subtle that is in the sense that men may be more likely to get research money than women just because of a gut reaction of a lot of reviewers, who are men because most of the scientists are men. So now this is purely conjecture, as you know. But what [was lacking] in that situation was actual support of me as a new faculty member trying to get established. In terms of being more nurturing and actually helping out; sitting down and having someone come around and say, "Well, why don't we talk about how you ought to do your first grant proposal. Let me give you some advice on how you might want to establish this laboratory, or how you might think about getting graduate students, or . . ."

Morse Δ There was *none* of that?

Rissler Δ No! No, no, no! I actually went outside the university, to the nearby US Department of Agriculture research facility, and actually developed a relationship there with another scientist, and did cooperative work and, in fact, worked in his laboratory some. No one sat down with me to talk about a strategy for getting started or anything like that.

Morse Δ Did anyone sit down with the young male professors?

Rissler Δ Now you see, I honestly don't know the answer to that specific question.

Morse Δ Even in a nonformal way? Did they go out and have beers?

Rissler Δ Oh, sure. There was a group of some young professors and some middle-level professors that for a time was going out drinking and partying and really relating to each other in a way that I found most unattractive. I'm sure there was networking and support in that. It's hard to know what is and is not going on. But I certainly had the sense that the men were, well, this is a hard thing for me, were more comfortable shall I say, than I was. In all honesty, some of that might be my own personality, being a bit reclusive, so it's a hard call.

Morse Δ Do science and technology represent "male" values? Weapons, pesticides, and your specialty, biotechnology—are these "male" things?

Rissler Δ Oh gosh, I think so. To me, well, there are a couple things that come to mind. I remember hearing a woman, a feminist woman give a talk; she was an environmentalist, and she was talking about men and responsibility, in science and in the home. She was sort of likening it, and I'm paraphrasing, "Who else but men would develop and promote a technology (nuclear technology) in which they knew not what to do with the waste." And it sort of reminded her of men taking care of children; somehow they don't; this is an older generation, they don't have to change the diapers, deal with the waste from children, from the household; that somehow they come in and they have no sense of responsibility for the whole life cycle of the child and the family. And I think that, I just can't imagine that a group of women scientists, given how women are socialized these days, would ever have developed and promoted a technology like nuclear technology, which we have no idea what to do with the waste and which is so dangerous. To me, I see this business of responsibility, and I almost see it in a sociobiological sense, that men in fact can abdicate responsibility for family and children and do that; and that women in fact cannot. And in fact they accept a responsibility for the entirety, for a larger picture than typically men do. I just see that in terms of child rearing, and I think it's broader.

One of the criticisms that we can legitimately make of scientists today is that they are unaccustomed to accepting responsibility for their work. You can see that in *Jurassic Park*; that scientists—and I've heard many scientists say this and, of course, they're mostly men—they say that "this is what I do in my lab, and I really can't control what happens to it beyond my lab." But in fact they could if there were a movement toward acceptance of responsibility. Now, to move on, I think that it is the same with pesticides in the sense that there has been no real acceptance of the responsibility for the impact of pesticides. We're putting out all these toxic chemicals, and the industry says, "Well, if you use the right equipment and the right protection, people are not going to get hurt. You just use this the right way, and you can sort of ignore that these are toxic chemicals." Well, that's baloney.

Morse Δ Down the road we'll see gender parity in the sciences. How will that change the entire outlook?

Rissler Δ Well, I'm not as optimistic as that 50% of women will be feminist women or women who are looking at things differently from men. One of the things that struck me a couple of years ago, when I was reading about how patriarchy is perpetuated, and this really struck me so deeply, is that women have in fact been used by patriarchy to perpetuate patriarchy, because they are in fact the teachers. That men have, in fact, employed women as the persons who teach patriarchy to generation after generation. And I participated in that. I taught for a number of years myself and I thought, "Oh my God, that is right. Men train women to perpetuate what they have." So, while I don't want to be so disparaging about women, I think a lot of us are just so socialized to accept what men think is best, and we just so internalize it that I think a lot of the women who are going to be in science are patriarchal types themselves.

Morse Δ So the Margaret Thatcher mode . . .

Rissler Δ Oh, yes, and I certainly see that. Look at Barbara McClintock. Look at how difficult her career was. And that's a real concrete way of thinking about how some women might do science differently. But I think it's real tough on women like [Barbara McClintock]. I mean, we get out rather than try to change the system from within. It is just too hard, I think, to try by yourself to change the system from within. But I think there will be some changes in science and how it's done, and it might just be in the environment, the atmosphere of science, because I think folks are just going to have to accommodate the women's still-pervasive desire and need to be the nurturers in the family. And I think this is going to change somewhat the politics of science. I don't know how it's going to change how science is done, and I don't know—it may change somewhat what we choose to do science on. I mean, why have we chosen, for example, to do so much reductionist biology versus holistic biology? I think it has a lot to do with men and reductionism; they've chosen this route. If women could really express themselves in terms of what we choose to do science in, I think we'd have a lot more support for holistic biology, for sustainable agriculture research. I think it might be reflected somewhat in the research priorities, if women, feminist women, can persevere to get up through the ladder to the point where they can decide what to do research on and how that money is granted.

Morse Δ What about from the other level in public policy, if we elect more women in Congress?

Rissler Δ I think that will be better, yes indeed, I do. I'm really pleased with some of the differences that one sees in terms of issues that are addressed in the Congress and as I watch on the local level in my state of Maryland. It's not enough, it's not enough change yet, but it's a harbinger I think, of what is to come. Now the question is, are courageous and strong women being attracted to stay in science like they are in politics? I don't know the answer to that.

Morse Δ There are plenty of strong women out there, including you.

Rissler Δ But in a sense, I got out. I didn't stay there to fight this battle.

Morse Δ You're not a bench scientist, but you're using what you've learned.

Rissler Δ Well, it is true. And it's interesting, if you look at the biotechnology debate, someone said to me, he wrote it in a letter to me just the other day, "If you look at who's objecting in biotechnology, really, at the federal level, it's three women and one man. And then if you look at who's doing the science, then it's mostly men. If you look in the industry, it's mostly men." So let's say ultimately, if and when we do have some affect on how the technology goes, it will be a lot of that different gender worldview, I think.

Congresswoman Marcy Kaptur, Democrat from Ohio, was sworn into her seventh term of office in January, 1995. She is one of only 57 women members of the 535 members of the 104th Congress. Kaptur was elected by her colleagues to serve on the powerful House Appropriations Committee, which is responsible for federal spending decisions. She is now the most senior woman on the committee. As part of her service on the Appropriations Committee, she sits on the Subcommittee on Veterans Affairs, Housing and Urban Development, and Independent Agencies, which has oversight on appropriations to the National Science Foundation. Congresswoman Kaptur's knowledge of and interest in scientific issues takes on particular significance based on her ability to advance scientific funding at the federal level.

Morse Δ Do you have a "woman's agenda" for the sciences?

Kaptur Δ One of the things we're involved in here on the appropriations side is to make sure there is funding for women in the sciences. My own deepest interest is on the education side, making sure that there is encouragement of women in the sciences. It's been wonderful, through the NSF and some of our other programs, to have young women from around our region called together and encouraged to go into the sciences by other women teaching the sciences, and bringing successful women scientists to their attention—usually when they're in the seventh and eighth grades, so they can think about science careers. "Science days," and this sort of thing, are pretty common in the area I represent in Ohio. They've made a real dent, I think, in getting women to participate.

Morse Δ Why do you think it's important for women to become scientists?

Kaptur Δ I think that a person should develop their own intellectual talent. We are all given some set of genes and abilities, and I just think people should develop their gifts, and those are equally distributed across men and women. I just think there should be equity there and that people should be encouraged. If they happen to be interested in it, I don't think women should be discouraged from studying in the sciences. I think that years ago that was the case in math and science. There's been a lot of attention given to teaching in this area now and to special programs that involve women. So I think that you're getting more and more women going into the sciences now, but the main reason is so people can fully develop their gifts.

Morse Δ I've been talking with a lot of scientists who are very concerned about the glut of science professionals, the oversupply of PhD-level scientists now. I'm wondering whether we should encourage more people to go into the sciences if there's not enough work for them.

Kaptur Δ Are you saying that there's not enough jobs for them, or that there are no jobs in teaching for them?

Morse Δ Very few permanent jobs across the board. It's a very tough job market.

Kaptur Δ Scientists? Well, that's amazing because they're so skilled. They're always telling us that the answer to America's future is to study more, and now you're telling me that on the science front that may not be true?

Morse Δ Yes. Quite a few people have told me that before we start encouraging more women to go into the sciences, or anyone for that matter, we've got to make sure that they'll have jobs on the other end. Otherwise we'll end up with a completely wasted resource. Is there something we can do at the national level to encourage the creation of more jobs for scientists?

Kaptur Δ Gosh, we've been putting so much more money into the National Science Foundation and into a lot of advanced technology programs. We, in times of budget cutting, have been increasing those dollars substantially. There are some projects that have been cut, like the superconducting supercollider, and the reason was simply budget. There's an advanced liquid nuclear reactor that's probably going to be on the chopping block, the space station is extremely controversial, and so I guess those aren't the only areas we need to be investing in. We have a lot of environmental cleanup to do; we have a lot of pharmaceutical research that needs to be done, a lot of agricultural research that needs to be done. I think it's in those areas that we'll see a lot of these wonderful minds move. Right now, I think the space station will probably be a go; even though I don't support manned space flight, I would very much support interplanetary exploration and unmanned space flight. I think money will continue to flow there.

Morse Δ Are there any "women's" science issues?

Kaptur Δ We in the women's caucus have been supporting an enormous amount of research that goes into cancer, arthritis, diseases that are known to affect women—osteoporosis—trying to get pharmaceutical research done so proper dosages are given. Women's health in general—using women in tests, this sort of thing. There isn't just one area, but several.

Morse △ What would the federal science agencies look like if they were all headed by women?

Kaptur △ I don't think I can answer that question specifically. I think that women would be making the most intelligent choices at the time, for where the nation should be making its scientific investments. They would meet the needs that are there, whatever those needs might be. I think that there would be a greater emphasis on health research, environmental research . . . but I just don't feel qualified to—I haven't studied how women have voted on this. I haven't studied the women's scientific community, saying, "Look, here's how we're different from the other scientific community . . ." I think that they would make choices in the nation's best interest. I would think that they would try to find ways to invest in defense that had dual use, as opposed to just military use. If you were to ask me as a woman, I'm interested in the productivity of this nation. I would ask myself what can we do, as the government of the United States, to catch up in so many of the fields where we're falling behind. I can answer it for myself, but I can't speak for women in general. I would look at the whole airplane business, automotive, the steel industry, the metals, the composites. I'm interested in the hardware that keeps the nation strong: industrially strong and agriculturally strong. I would be interested in promoting better transportation systems, new fuels, new metals, composites, and helping our industry meet the competition from Germany and Japan. Because if we don't do it, our standard of living will continue to sputter. That's where my own personal interests are.

Morse △ So it sounds like you are very interested in R&D.

Kaptur △ Oh yes, oh yes.

Morse △ Do you ever encounter sexism, or are you ever discounted when you're discussing science or technology policy?

Kaptur △ Well, I think that when you get elected here, you have to prove yourself to your colleagues. It doesn't happen in a year or 2 years. I think once you gain respect, and I can't tell you all of the pieces of that, you are listened to. You have to develop expertise on an issue and you have to be the best. At that point, you gain respect. There will always be those that make comments that are negative toward people, but that happens in general society. And I think in some ways, because people up here have to run for office and they have to pass a certain level of societal acceptability in order to be here, they're probably more tolerant, they're broader-thinking people. So we probably don't have quite as much [sexism] as exists in other sectors, although you always do have some; but I think basically, it's a pretty good environment in which to work.

Morse Δ Any parting words?

Kaptur Δ I think we have come through an enormous revolution in this country in the last 25 years, of women moving into fields of endeavor where they never had before. In all of our schools across the country, we are seeing now the advantage of involving women, and I think that America's going to make it in the 21st century because we have unbottled this tremendous resource that was locked up for such a very long time: to allow women to use their minds to the extent that they wish. The one piece that we haven't been successful in yet is how do we handle family life at the same time as people pursue their own academic and intellectual interests. We are suffering all across this nation from family structures and communities that are falling apart, because we haven't addressed the new world of work. That is something that government can't solve. It has to be solved inside the family, in each company, in each work location, in each school system . . . we have to adapt our institutions to this new life. And we haven't.

7

A Brave New World
Women Speak on the Future of Science

Throughout this book there has been a thread, woven among the interviews, survey responses, and commentary, that all is not quite as it should be in the sciences. Each respondent defined the problem in a different way, which is certainly to be expected when talking with such a diverse group. Women have taken issue with a variety of science's characteristics, from things as obvious as the continued underrepresentation of women and minorities, to more controversial qualities, such as expectations of untenable working hours and a lack of openness to scientific questions outside of the accepted mainstream. The men I've spoken with have been in the main quite supportive of the women's positions, but in both groups a considerable amount of resignation to the status quo remains.

In this chapter, I asked established women scientists to speak on the matter of the future. The women in this chapter have witnessed, and in some cases spearheaded, the recent historical and political movements that allowed women to enter the sciences at all. This experience gives them a perspective that younger women cannot have, and enriches the value of their statements.

The scientists in this chapter are identified by name; they are, we hope, beyond censure for their opinions.

Dr. Maxine Singer received her PhD degree in biochemistry in 1957 from Yale University. In 1988, following a distinguished career in genetics and cancer research, she became president of the prestigious Carnegie Institution of Washington, DC, a heavily endowed independent, nonprofit institution for research and education, whose five research centers carry out basic research in the physical and biological sciences. Dr. Singer is a member of the National Academy of Sciences, and in 1992, she received the National Medal of Science, the nation's highest scientific honor bestowed by the President of the United States for her "outstanding scientific accomplishments and her deep concern for the societal responsibility of the scientist." I spoke with Dr. Singer from her office.

Morse Δ What do you think science will look like in 10 or 20 years?

Singer Δ I don't really make predictions about science, because predictions about science usually turn out to be wrong. The things that are most exciting are always the ones that nobody's really thought of. So it's extremely difficult, except for kind of stating the obvious, with respect toward kinds of things that will happen in different scientific areas in the next couple of years.

Morse Δ This is not so much about the findings or the discoveries that will occur, but more the cultural changes that might occur. The women I've been speaking with have told me some of their frustrations with the field, as people who are still in charge of raising families—trying to get tenure, trying to break into a field in which there are very few jobs right now. I'm asking them how they're going to deal with that, and whether they think young women should be encouraged or discouraged from entering the sciences right now if there really isn't a professional path for them once they receive their PhDs.

Singer Δ Well, actually, I recognize all of those issues. They are particularly close to home, since one of my daughters is in fact starting such a career and has a new baby. We've been talking about a lot of these things recently. She's been sharing her experiences with me. But one of the conclusions that we've come to, one of the problems, both for women and for men, is that their training does not really instruct them in the range of things that one can do as a scientist. And this, I believe, is emerging as one of the most important issues to be discussed in the next couple of years. Science professors, by and large, go on the almost-unspoken assumption that their PhD students will go on in an academic or scholarly track of some sort. And more than that, the message is conveyed both implicitly and explicitly that that is not only what is expected, but that it is what's considered the highest good. There's a growing awareness that it's not doing anybody any good—women or men—and that in fact there are many different kinds of opportunities for people trained in science to do interesting things all their lives and their research training prepares them for [those jobs]. In the most enlightened view, that even includes teaching in high school. But it includes an enormous range of things: for mathematicians, it includes the world of finance, it includes the world of manufacturing and industry; for scientists and mathematicians and engineers in general, it includes areas of science policy, it includes public education, it includes all kinds of regulatory situations; there is just an enormous range of things that aren't even in the vision of many young people now finishing degrees. And as I say, because now there's an assumption that those things are not quite as good as other things, people, even if they learn about such opportunities, always think about them as somehow second-best or second-class. There's going to be a report issued this fall by the National Academy of Sciences, with respect to advanced education in science, which is going to make this point very, very strongly, and make very specific recommendations for the kinds of changes that ought to occur in graduate education in the sciences in order to try to open up for people all of the opportunities that there are.

Morse △ Is this a response to the current bad situation, or is this a prescription for all of the future graduates in science? Are we responding to a temporary situation, and then hoping that fewer PhDs will be graduated?

Singer △ This is not to say that fewer PhDs will be graduated. In fact, the thrust of the report is the opposite. Not only do we not need fewer people, we need more people. But we need to realize what this training prepares people for and how the society can use them. The other side of the coin is the kinds of jobs such people can find. I think, like all such things, there was probably a stimulus to looking at this question that came from the job market, but the nature of the draft of the report I saw is much more general and far-reaching than just dealing with this year's job situation.

Morse △ I have a lot of feeling about the high school teaching option. In talking with scientists, that's a red flag for them. If they were to go through 8 or 9 years of graduate training, and then get out and be paid $25,000 per year to teach high school, then . . .

Singer Δ That's right, of course. It requires, well, first of all, high school science teachers earn a good bit more than $25,000 a year, in most places right now.

Morse Δ Right, but I'm thinking of a rural Minnesota high school teacher.

Singer Δ Yes, but science teachers, even in rural Minnesota, they have salaries to attract people. But still, salaries are an issue, and that will have to be dealt with. All these things are problems and issues to work on. You know, there are people even getting interested in elementary school education, and working on curricula, and so forth. Not teaching, because they're really not trained to teach, and they wouldn't necessarily be good teachers; but they are working on curriculum, they are working on training teachers, there's a whole variety of things being done, and more and more people are actually working in these areas.

Morse Δ So really, as a nation, we have to be more creative about how we use this brain trust.

Singer Δ Well, you could put it that way if you want to, but really, these people are needed. There are many projects going on all over the country, in state governments and in universities, that deal with education issues. But also, there's the whole world of industry. One of the kinds of things that people have considered is that graduate training should include, as a requirement almost, to spend some time in an industrial lab, so that people would learn what it was like to work in such a lab. Of course, in biology, increasing numbers of young PhDs are going into industry because of all the new biotechnology companies. Those companies have made 80,000 jobs in the last 10 years. That's not trivial for PhDs. So that's an example of what's going on. It's my understanding that many young women are finding those positions really very satisfying. The research is interesting, the colleagues are smart. So there is a whole range of things that can be done, and we need to inform the students about these possibilities, and we need to make it clear to them that these are perfectly respectable, if not desirable, things to do.

Morse Δ May I ask you, on a microscale, about how women can reasonably have a scientific career, and a family, and a community life?

Singer Δ Well, I'm not sure that you can have everything. In fact, I'm sure you can't have everything. And you have to make some choices. If you choose to have an energetic career in science, by which I guess I mean the kind of career where you have a high degree of independence in what you do, be it in industry or academia, or in government, or whatever, and have a family, then probably there's not much else you can do. So if you're very interested in community work, or political activities, and so forth, probably one or the others would have to go. You can't really have everything; nobody can. That doesn't distinguish scientists from anybody else. Everybody makes choices. I think it's feasible, but of course I'm very biased because I did it, and I did it at a time when there was no assistance of any kind. Now, at least, there's some modicum of assistance in different places. I thought that at the institution where my daughter is an assistant professor, that while there was a certain amount of struggle involved in it, it was not untoward, and in the end they were quite sympathetic and even understanding and generous in the arrangement they made for her. I think that's increasingly the experience.

Things are changing. I think many of the men are very sympathetic with the changes, and many of them absolutely have their heads in the sand; they can't be moved. We'll just have to wait until they retire and die. But there are a lot of men, and not necessarily all young ones—in fact, I'm not sure it's age related at all—who are willing to go to bat for arrangements that make it somewhat easier for women to manage. There are, nevertheless, and I think it's really important to recognize this, there are inherent problems that nobody can do anything about. There are institutional matters, which can be dealt with, even though they may mean a fight. But there are matters inherent in fields that probably can't be dealt with. The pace of science, for example. It just moves very quickly; more quickly in some fields than others, obviously, but it does move very, very quickly. And half-time or a year out is a major stumbling block. And it's very difficult to deal with. Some young women complain about that as being unfair. That's a useless complaint, because that's just the way it is. It wasn't designed to be unfair, like some institutional matters are. So you have to find a way to deal with that. I think, for young women, they need to find a way to deal with that.

Another issue, which I find comes up sometimes, [is that] economic, or I should say lifestyle expectations, are relatively high in the young generation. I find often in talking with young women who come to me to talk about their particular issues, particularly people who have waited to be married until they're older or to have children when they're older, that they've gotten accustomed to a certain lifestyle. They're not quite willing to be paupers again. Those in my generation who had no help, that is to say no institutional mechanisms for making things easier in any way, we just sort of assumed that whatever the family's income, a very high percentage of it would be spent in child care, and so forth. It was an era when we were raised during the Depression, we were coming out of the war, we didn't expect to go out for dinner, to buy clothes, and so forth. I think young people who have grown up in an affluent condition in the last 20 years have a somewhat different approach. So they look at me, when I suggest that maybe they could spend more money on this, and they say, "Well, then I won't be able to do X, Y, and Z!" And you say, "Yes, that's the choice you have to make." Again, it comes down to the fact that you can't have everything. You do have to make choices. I don't mean to imply that everybody's rolling in money; I don't. But I hear this.

The other thing I hear in such conversations is always so interesting to me, and it may be interesting to you. A young woman will say, "But I have to spend all my salary on child care." And I look at them and say, "Your salary? What about your husband's salary?" And so there's still this lingering notion that children are the woman's responsibility—even among women who would in every other way be talking differently. They'd be talking differently about sharing chores, and time commitments, and so forth, but when it comes to the money, suddenly it emerges. So you know that they're still thinking that way. They're really taken by surprise when you challenge that remark.

Morse Δ One of the interesting points that's come across to me is that not only are people loathe to pay for child care, but we don't want to pay child-care providers anything approaching a livable salary.

Singer Δ Bad news. I had all the help I could get. That was my personal approach. Of course, I had a lot of children in very quick order. There wasn't day care available, but I wouldn't have used it anyway, I had so many children at home. I had in-home help for the children full-time, I had somebody who came a couple of days a week to clean, and we always had a student living with us so we had an extra pair of hands and a built-in baby-sitter. A lot of money went, and nobody seems the worse for it—in fact, everybody seems fine. They're all young adults now. And my new granddaughter just began day care this week—aged 3 months. I think it's going to be fine.

Morse Δ Can you even approach giving me a snapshot of what in your mind science would look like if half of the scientists were women?

Singer Δ You mean, what *science* would look like? You know, it's really important to make this distinction.

Morse Δ Some people do say that science itself would look quite different.

Singer Δ Why would science look different? Everybody looks at science differently. Every single individual. That's what makes it such an interesting endeavor. And women will look at it differently, but I'm not sure that the range will be any different. Do you see what I mean? The kind of disciplined thinking has to be the same or it will be trash. The imaginations will be in somewhat different directions, but so will men's. I have trouble with that one. It may be that I am quite wrong and I understand that. But I have trouble with it.

I know a lot of women scientists. Most of my really good friends are scientists. I don't see that the range of ways that people look at things, or imaginations, are different. I think a lot of the stuff that's been written by extreme philosophical feminists about science is really a lot of garbage. Science will keep on trying to explain the natural world, and if people have a different agenda, then it's not science. Most of that writing is about culture, not about science. You'd have to believe that the culture in which science is done makes a difference to the science to buy it, and there's no evidence for that. I mean, you look at great scientific discoveries worldwide, made in different cultures, and they turn out to depend on the same general requirements for rigor and testing, and so forth. And yet they come from very different cultures. I've tried very hard to read [the feminist literature] and I find it extremely difficult, because it's not about science.

There are certain things about science which have reached a level of certainty that's very high. For example, the structure of DNA. There are still fine points to be learned about it, but it's clear that our current picture of the structure of DNA must be pretty close to what it is or we wouldn't be able to do most of the things that people do in molecular biology these days. So to talk about such constructs as somehow being culturally biased doesn't make any sense to me. And usually what it turns out is that the organization of science or the culture of science or something is culturally biased. This doesn't mean that what's learned is culturally biased. And that's what is so often missed.

Morse Δ Do you love your field?

Singer Δ Oh yeah. (Laughs) Oh yeah. No question. I had a bunch of lucky circumstances, and to the extent that I made choices, I made the right ones for me. I've never been sorry for a minute.

Dr. Anita Borg is a 45-year-old software engineer at Digital Equipment Corporation in Palo Alto, California. She is founder and administrator of the electronic discussion group "Systers," created as a communications tool for women doing technical work in computing. Systers provides long-distance mentoring and information exchange among women who are frequently geographically isolated from one another. In 1994, she received the World of Today and Tomorrow Award from the Girl Scouts of Santa Clara County, and was named one of the top 100 women in computing by *Open Computing* magazine. I spoke with Dr. Borg from her office on a Sunday afternoon, where she had arrived to do some work after completing six landings during her flying lesson.

Morse Δ Can you tell me about your career?

Borg Δ Sure. It started out in kind of a weird way. I got married very young, and wound up going to New York with my husband, and quitting school for 2 years in the middle of my undergraduate work. I needed a job. I had been a math major—this was in 1969—and so I got a job as a girl Friday to the data processing department of a small insurance company. I taught myself to program, and when I finally got divorced and got the heck out of that awful job, I decided that programming was pretty interesting and I went back to school at New York University as a computer science major. I stayed at NYU for 10 years, through my PhD. I did a whole lot of other things on the side— one of the reasons it took so long is that I was never willing to do the "nose-to-the-grindstone, don't do anything else but your graduate work because that's the only way to get a PhD" routine. I renovated houses, and I renovated apartments, and paddled kayaks, and rode motorcycles, and did hiking trips, and went to Europe and all of that. I had a real life.

Morse Δ So you're an oddity in the field.

Borg Δ [Laughs] It's always been important to me to have a real life. The next thing I did after my PhD, well, first of all, I took 6 months off in between, and then I went to work for a start-up in New Jersey call Auragen, which was building a fault-tolerant operating system. I was about the 11th employee. It was a great opportunity, although I nearly killed myself for about 3 years. This is the point at which my ideas of having a balanced life went out the window.

Morse Δ Was that because it was a start-up company?

Borg △ Oh, yeah. Start-ups are deadly. And, it was my first job and I was fairly insecure. I didn't realize until the company was starting to get smaller instead of larger that this wasn't my fault, and that it was important for me to take vacations and not work 12 hours a day. So I worked there for 3 years, and when they were heading under, Nixdorf in West Germany bought the technology. In the middle of that, I wrote a paper about what we were doing and it was accepted for a conference. That was my *real* introduction into the world of research, because my graduate career had been a little bit isolating. I met lots of people there, and about a year later, the folks from Nixdorf hired me away. As a result of being one of our funders, they owned the technology and they hired me to come and finish some stuff up. So I went, me and my two cats went to Germany for a year. It was a character-building experience! Going from living in Manhattan for 15 years to living in a small town in northern Germany, and not to speak German, was an interesting experience. But then I took a 6-month break after that and traveled around Europe a while. As a result of the contacts I made with my first paper and another paper that I gave right at the end of that, I met people at Digital and got my job at the Western Research Lab. I was hired to do operating systems stuff, but the project that I was going to work on wound up not happening, so I twiddled my thumbs for a little while and came up with another project, which was much more connected to computer architecture and figuring out how memory systems work. That took off, and I got some really good work done and did some fairly ground-breaking work in that area.

While I was in the middle of that project, I started the Systers mailing list, which started out as 20 women in operating systems who were going to talk to each other and is now 1780 women from all technical aspects of computing. As the list got really big, I got unhappy with how much work it was taking to administer it, and standard broadcast electronic mail didn't offer the kind of communication that people wanted to do; so, I started thinking about that and came up with some ideas, and lo-and-behold, I wound up in a completely different research area. Now, I'm building e-mail communication tools, as a result of having my feminism on one side and my computer science on the other. Finally, after nearly 25 years, they're merged together and allowing me to do both, which is really tremendous fun. To have a research project grow out of this sideline mailing list was wonderful. I think having your vocation and your avocation come together is where you can actually do the most, because you care about what you're doing. I probably care more about the technical work that I'm doing now than I ever have. It was always interesting, I always had a great time doing it, but the goal was always, well, make computers faster or make them not crash. Make computers do something that business wants them to do as opposed to make computers do something that *I* think they ought to do.

Morse Δ How long do you think that this cohesive moment will last?

Borg Δ I'm intending to make it last for as much of the rest of my career as I can. It's real important to me, and I have a whole lot of ideas that aren't as directly related to feminism but to the whole issue of what it means for people to communicate effectively in a group that's very, very diverse. It's very interesting. We have a very diverse population on all sorts of different dimensions, and I think that a great deal of the work that's going on isn't really taking into account what it's going to mean for all sorts of different kinds of people to be communicating with this new technology. I'm also very concerned with issues of universal access and making sure that there are provisions for access for all sorts of different people, rather than the standard business sorts and people who have money. It's tricky to focus on that, because we do live in a capitalist society, so you have to figure out how somebody is going to make money doing this or how it's going to get paid for. One of the things I'm thinking about is that it's critically important not to just have technologists looking at this. I would love to have a group of people that included futurists, and economists, and sociologists, and anthropologists, and technologists; talking about communication and understanding how to both build the right stuff, and introduce it, and make it meet people's needs, as opposed to building something and then kind of forcing it down their throats and making them need it.

Morse Δ Do you think it's true that a lot of technical people are not allowed to let the other parts of their consciousness and sensibilities out through their work?

Borg Δ In part, that's true. In part, corporations don't really allow people to do that. On the other hand, I think that people in computing frequently have the direction of our education, the direction of the values system in the computing field, which is, "I can make this neat thing happen; let's figure out how we can use it." It's cool to make computers faster and this is a neat technical idea; now what are we going to do with it? But industry is changing rather dramatically at the moment, and I don't think it can continue to exist directed by engineers. And most people think that's true—the computer industry is now moving in a direction where we really have to be paying attention. The standard thing is to talk about it as being a commodities market. But in fact what that means is that you really have to look at what people want, and what people can use, and what's going to help people, and then develop *that* as opposed to [what was done] in the early stages, when this was all brand new: it was just build whatever you can build, explore the entire space. There's room for some of that research, but considerably less. I think that's a good thing. I think it's really good for technologists to pay attention to all sorts of implications of what they do. And I really do believe that the women in the field have a tendency to do more of that, to pay more attention to what the implications of what they are doing are.

Morse Δ Any idea why?

Borg Δ It's hard to come out with an answer that isn't just, "It's obvious— we're women!" I think that we just have a tendency to have a broader picture of life. We're always doing this family and career balancing, life and career balancing. This notion of balance, this notion of "How does all of this fit into my life?" Well, how does all of this fit into everything else? We're also, just generally, more revolutionary. Anybody who's in computing or science at the moment has had to learn to challenge all sorts of different ways of thinking, about what's appropriate for women, about what you thought was appropriate for yourself. Once you've done that a lot, it becomes habit. It's probably easier for women, in a way, to challenge standard ways of thinking about things, because we've been doing that in order to get here to begin with. One of the problems, of course, is that we're not quite comfortable with the fact that that's a really cool thing to do. You're forced to break all these molds and challenge all sorts of stereotypes, and there's a tendency to say, "Oh, God, I'd really just like to fit in for a while." On the other hand, what we can do is come together and get a lot of mutual support and realize that that is an incredible tool, that ability to challenge things and look at things from a different perspective. Bringing the female concern into science and into the technical fields is really good for them. Now, it's incredibly difficult. I was talking to a woman who just recently left Xerox, because she found that working within a large company, working always by male rules, was just too much. She was just sick of it. She managed a group of 15 people and was quite successful, and just got sick of it. She didn't want to fight those battles anymore. What we talked about was that, until we really do have critical mass of women in these companies, that in spite of the fact that we'd like to see things done differently, they're not really going to change. We can change things in little bits and pieces, but not in whole cloth, which is why it's so terribly important to get more and more young women interested in science and technology and into the field, and to support them and have support structures all the way from K through 12, and up through college, and in your early career, so that there's never a point at which you feel completely lost. That simply doesn't exist right now.

There was a wonderful conference in 1993, supported by the National Science Foundation, for undergraduate and graduate students in computing, one from each school, so it was a really diverse group. I told them all about Systers and the fact that there is a community of technical women out there who support each other. And this one young woman came up to me in tears. She's a graduate student at a midwestern university, she's the only female in her department, there's no female faculty, she has never known any other women who were seriously involved in computing, her family doesn't understand, and she was about to give up. And all of a sudden, there's this huge, huge community that's out there and she's going to stick it out. There are all sorts of programs out there and they're all scattered. One of my visions is that I'm going to find a way to get all of these things connected. I do networking research, and what good is that if I can't find out how to connect up all these programs, for people of all different ages, to help them so that there is this continuous support for anybody who has interest and aptitude to stick with it. So then I think that the long-term goal has to be to get more diversity at all levels of science and technology, and to bring concerns about how it gets used, and why we're using it, besides just technology for technology's sake or technology for money's sake. There are lots of good reasons to use technology. I'm not a Luddite by any stretch of the imagination. But I think all too often we don't challenge existing ideas about how you can make money, for example. Or what is interesting research? I would argue that developing all the different kinds of hardware and software that are necessary to connect poor people and disabled people, which probably have to be small, and cheap, and useful, I think there are just as many fascinating challenges in doing that right as there are in making the expensive stuff that allows hospitals to get all their data back and forth to each other, which requires lots of equipment, and lots of bandwidth, and lots of fancy stuff. They're both really interesting challenges, but it seems to me traditionally, the latter is the one that is touted as the interesting problem.

Morse Δ Could you address the process of developing technologies in terms of working cultures? Can people work fewer hours?

Borg Δ I don't know about the hours business. The amount of hours that I work varies all over the map. When I'm having a low point, and I'm not feeling terribly creative, and I've got other stuff going on, I might work 30 to 35 hours a week. When I'm really into what I'm doing and I'm having a great time with it, I might work 50, 55 hours a week. But you know I must admit, I'm in a rather privileged position to be able to do that at all. The number of hours I spend on things is much more defined by me than anything else. As far as work culture goes, there certainly is, in science and technology, a macho kind of value placed on being completely dedicated and devoted to the technology. I find that a lot of women think that that makes it very hard for them to be taken seriously, because they have other interests. They try to have a balanced life. I certainly believe that, as I said when I was in graduate school, I refused to be totally devoted and dedicated to computers. And I've gone on to make probably more contributions in a lot of different fields than many of my colleagues who never picked their heads up and looked around.

Morse Δ Are you gifted? I'm told that there are those scientists who are so talented at what they do, that they don't need to put in the kind of grueling hours.

Borg Δ Well, you know, I'm very smart. But I've never viewed myself as being one of those geniuses. And I'm not even convinced that those geniuses always have the best ideas. I think that people have really good ideas when they are in an environment that nurtures their interests. So, for example, women stuck in an environment where the only model for communication and success is incredibly hierarchical and incredibly competitive are not in an environment where their potential can necessarily show through, because they are struggling with an uncomfortable environment. I wasn't really sure about this until I held the Grace Hopper Celebration of Women and Computing conference. As the result of my experience in the field, and having read Deborah Tannen's stuff on communication, was that if women get together, and if we have this critical mass, then will communication be different or not? Clearly, most of us have gotten where we have because, to some extent, we've been able to work the system as it is. I can be very aggressive. I can be very competitive if I need to be. It's not particularly the way I like to work, but I'm successful because I've figured out how to do that, how to work in a foreign environment.

So we had this conference. A conference plays a couple of different roles. On the one hand, it's a place where people go to see their role models. They go to hear the senior people talk, and to see what success is like. On the other hand, they go there to generally communicate with people in their field. A standard conference in computing is, if you're lucky, 10% female. The model of communication is very competitive, there's a whole lot of ego, a whole lot of posturing going on, a whole lot of positioning oneself in the hierarchy. This conference was 450 technical women in computing, getting together to listen to 16 of the absolute top people in computing who happen to be female. It was completely and totally different. The attitude of the speakers was not posturing. The attitude of the speakers was, "Here's what I'm doing. I really love what I'm doing. Can I get you excited about this? Would you like to do this too?" They were really sharing what they were doing and trying to get people excited about it. The communication outside, both in the formal question-and-answer periods and out in the hall, was very upbeat. The graduate students were absolutely thrilled. They went to a conference and they could talk about what they did and not feel like the first thing that was going to happen was that somebody was going to shoot it full of holes. There's this tendency to see, to look for, the positive stuff in what you're hearing, not to be blind to the problems, but to support the positive stuff, which is a completely different kind of communication about technology, that you just don't see happen except in small groups, where people are very comfortable with each other and they don't have to play those games—and they still play them a little bit! This kind of conference was proof to me that there really is a different way of communicating, that my gut feeling about this was not wrong, and that it makes it even more important, for the future of society, that we bring these other kinds of communication and other ways of thinking about problems into the mainstream. I know one doesn't have to win. I hope.

Unfortunately, it's like someone said: Comparing the male and female styles of communication is like comparing the New York and California styles of driving. When the New Yorkers get to California, it's the aggressive style, and then California driving turns into New York driving. There may be some inherent problem with that, but we're never going to find out about that unless we get lots of women in the field, saying, "Look, I don't like the way this works," and have a bunch of other women saying, "Right, neither do I. Let's do this differently." We need women running groups who say, "Look, shooting people full of holes is not the only way to analyze a technical idea." There are ways of looking for the positive without being blinded by the stuff you don't like. In my opinion, that is a much more positive, productive, good for science, good for creativity, way of doing things.

Morse Δ How many men do you think feel that, but are reluctant to speak up?

Borg Δ You know, it's very hard to know. I think a lot of them would say, "Yes, I agree with you," and practice it more or less in small groups. Some would simply say, "Get real, that's not the way the world works."

Morse Δ For them.

Borg Δ Yes. But, you know, what is the world? The world of technology is 90% male. The "real world" is the male way of doing things. I run Systers as a very supportive kind of communication environment. There are rules and etiquette laid out for the appropriate way to interact on that list. If people don't like it, that's their problem. Most of the women on the list are extremely happy with the way I run it, but every now and again I get a message that says, "Hey, get real. This isn't the way the Internet really works, this isn't the way people communicate on the Net." And then I say, "Well, that's all good and well, but I've never been one to follow rules, and I don't intend to continue just doing stuff the way everyone else does it." If I weren't breaking new ground, I wouldn't be happy.

Dr. Evelynn Hammonds is assistant professor of the history of science, in the Science, Technology, and Society program at the Massachusetts Institute of Technology. Dr. Hammonds has written and spoken widely on the issue of gender and race in science, including articles and book reviews on AIDS, the feminist study of gender and science, and health care. She was a coconvener for the national conference "Black Women in the Academy: Defending Our Name, 1894–1994," held at MIT. Her doctorate is in the history of science, but she also has undergraduate degrees in physics and electrical engineering, and a master's degree in physics.

Morse Δ Why did you decide to go into those fields?

Hammonds Δ I started as an undergraduate studying physics and engineering because I have had a long interest in science since I was a child. I went to a joint program at Spelman College and Georgia Tech, a dual-degree program. I studied physics at Spelman and electrical engineering at Georgia Tech.

Morse Δ Did you have any role models when you were a child? Or did you just enjoy the subject so much?

Hammonds Δ Well, I enjoyed the subject, but my father was a big influence. He had a long-standing interest in science and engineering, but growing up in Georgia, there were no engineering schools for African Americans. During my father's college years, Georgia Tech was not open to African Americans, and so he wasn't able to pursue his interest in it in the ways that he wanted to. But he maintained a long interest in it and encouraged my own interest in it, so I had a lot of support from my father and from my mother as well. Both my parents were big influences on me in that respect.

Morse Δ Were there a lot of women of color working with you? Did you find that, as women of color, you had more barriers?

Hammonds Δ I entered college in 1971, and Spelman College is a black women's college, so there were other black women there studying physics with me. When I transferred to Georgia Tech and electrical engineering, I was the only African American woman in my class, and there were only two other African Americans besides me. I encountered in my science and technical education, particularly when I transferred to white schools, both racism and sexism, to the extent that I had professors or fellow students who didn't think that I should be studying a science either because I was black or because I was a woman.

Morse Δ A double whammy.

Hammonds Δ Absolutely. So that, I think, was psychologically difficult, and sometimes, just practically difficult. I mean to work in lab groups—when I was at Georgia Tech one of the experiences I had over and over again was to go to lab and find that male students would want me to take notes on the lab, but not do the actual work of the lab. They'd say, "Well here, take a pad, take down the data," and they would do all the setup and run the experiment.

Morse Δ So you were the group secretary.

Hammonds Δ Yeah, right. I always had to fight to have an opportunity to actually do the work. And I often didn't get support from lab instructors or course instructors for that either. They didn't think there was anything wrong. They were like, "Well that's a good deal, you don't have to do . . ." So you know, the general attitude, certainly at Georgia Tech, was that women were not taken seriously, and it certainly was absolutely true for me. I think some people had a somewhat different experience, but I think one could safely say overall it was very much that. I did my master's degree at MIT where I was also, for much of the time I was here, the only African-American woman in my class and in my department. At MIT I think there were different kinds of barriers. It's a very difficult, demanding school, but there certainly weren't a lot of women around, and there certainly weren't a lot of women in my field, and there certainly weren't any other African-American women at the institute at that time in my field. So [at MIT], rather than sort of more overt, sexist comments, I felt pretty isolated, and really had to deal with certain kinds of isolation.

Morse Δ How did you do that?

Hammonds Δ Gee, I don't know. I think I tried to find other community, I tried to find other outlets, where I would feel more a part of a community. But I began to perceive what was in store for me if I were to finish my PhD and really be a physicist. The isolation wasn't going to go away. That was just the nature of the situation at that particular time, that there just weren't a lot of women doing it. I took a leave of absence and really wanted to think about whether or not I wanted to keep doing science in the lab for that reason, and at the end of my leave I actually made the decision to enter the graduate program in the history of science at Harvard, because I felt that I would still have an opportunity to think about science in an analytical and rigorous way, but the sort of social dynamic, that kind of isolation I experienced in the laboratory, I wouldn't have to deal with.

Morse Δ Do you miss bench science?

Hammonds Δ Yes, I do miss it, actually. There are things I miss about it a great deal.

Morse Δ Do you feel strongly enough about it that you would say you were deprived? Because of the hardships, that you were deprived of your initial, your original choice to be a lab scientist?

Hammonds Δ No, I don't think "deprived" is an appropriate word. I mean, I chose to not do it. I think at the time that there were a lot of people, when I chose not to complete my PhD at MIT, there were certainly a lot of people who were very disappointed and who tried to assure me that people wanted me to do it and that there was a lot of support. It was my own choice, and I don't feel that I was deprived. Having the opportunity that I have in history of science to think about science in its broadest terms, in this particular context, is something you don't have the luxury of doing while you're actually doing it.

And I like that part very much. For me, history of science sort of brings everything together, except you don't get to do lab work.

Morse Δ I imagine a lot of people wait until they retire to think about the larger picture, and then regret not thinking about it when they were younger.

Hammonds Δ Precisely. When you have a working lab, you just don't have time to do anything but get that work done.

Morse Δ Do you think, in terms of "having a life," that the nature of science would have to, or could change? Is it true that in the United States women must choose between a science career and a life?

Hammonds Δ I think that choice is a real one in science, particularly if you're an experimentalist. You're competing with men who are able to be relieved of their family responsibilities in order to devote 99% of their time to their science. Those are the people you compete against, and that's what you have to do, unless you're incredibly smart, and most people aren't that smart. It's rare people who can put in less time and get the same quality of work done. So I do think that it's a major hindrance for women who want to have a family. There are beginning to be structures put in place to help people do that. For example, [there are] professorships which provide, [in addition to] research money for young women scientists, some setup funds for child care. I mean, it's just a necessity, as long as we have a culture where women are primarily responsible for child care, particularly for young children. The issue is just not going to go away. For women scientists, it's critical. The only way you can do it is if you have some help.

Morse Δ Would women scientists, who obviously must be strong and smart just to become scientists, have major clout if they got together en masse and said, for example, "We're going to change the workday"?

Hammonds Δ I don't think that's going to be possible until there are significantly more women scientists with clout; that is, women scientists in the senior ranks, numbers of women scientists who have successful laboratories, where they have lots of graduate students and postdocs working for them. That's just not the case yet. There are still, here at MIT, very few women in the senior ranks. It's going to take a transformation in terms of numbers of women in the senior ranks, and also numbers of men in the senior ranks who also begin to want to restructure how scientific work is done. And that may take another generation. I would suspect it is going to take another generation.

Morse Δ What about the ethics of science? Do women approach ethical decisions differently?

Hammonds Δ Well, I think that question is an old one. One has to be careful of generalizing about women. Women who become scientists are trained very much in the same ways that men scientists are trained. The focus is less on gender and more on being a particular kind of scientist. I don't think many scientists get clarity about the specific ways in which they do their own research until they are fairly mature in that work, and mature and successful enough to take particular kinds of risks to do their work is specific kinds of ways. So I don't think there's anything special that women bring to their scientific work given their training that would lead me to support, at this particular historical moment, the notion that women will do science differently. I think there are some women who come to their scientific work, because of their own political, professional perspectives, who want to ask certain kinds of questions that other people in their field have not asked, and I think that is an issue that's much more prevalent in the biological and life sciences and relevant in the biological and life sciences than it is in the natural and physical sciences. So for chemists, for physicists, for even molecular biologists, which is becoming a lot more like chemistry and physics, I think it's very difficult to ascertain a particular kind of style of work for anybody. I think the sort of gross generalization that women will perhaps do science differently is a problematic assertion, and you have to get very specific about it.

For a generation of women scientists, I think particularly many of the ones who are senior now, that question is a very troublesome one, because they came up at a time when they were trying to prove that it was possible for women to really do good science. And then for people to now raise the question about women doing it differently, well, women wanted the opportunity to do science the best possible way that anyone does science. And that's part of the aspect of science that is sort of universalizing—a work that has certain universal characteristics rather than particular characteristics based on who the researcher is. That is, I think, a very strong and deeply held belief among scientists and among women scientists as well as men. There's a younger generation which is raising that question, and in the feminist studies of science that question has been raised. I think it's a valid question to raise, but I think it's valid to raise on one set of discussions; but for actual practicing women scientists, [given] what they have to do in their lab every day, I think the question, especially in fields like physics and chemistry, is not a particularly relevant one at this point in time. It may turn out to be relevant later, but not at this point in time.

Morse Δ Could it be that scientists get blamed for things that might really be in the realm of policy—such as locating a pesticide factory in a low-income neighborhood, or the kind of environmental racism that happens with hazardous waste technology?

Hammonds Δ Well, I don't make big separations about science and policy. I think there are scientists involved in decisions about where to locate plants like that, or scientists involved in research on toxic chemicals who I actually think should take a greater role in discussions about how toxic chemicals will be disposed of and where they will be disposed of. The record probably shows more scientists on the side of the industries involved than on the side of the communities where these chemicals are having a detrimental impact. I think that scientists should be, not necessarily blamed, but held more accountable when they are involved in such projects.

Morse Δ Do you think that scientists believe that responsibility for the outcomes of their work is beyond the boundaries of their laboratories?

Hammonds Δ I think some scientists certainly feel that. They would rather not address the application issues. I don't think in this day and age, [considering] how people get funding, that it's really possible for people to do that easily.

Morse Δ So the funders are going to have a lot of control over what happens.

Hammonds Δ Yes, I think so. I think that's always been the case. I mean, funders help set scientific agendas.

Morse Δ Do you think there will ever be gender parity in the sciences?

Hammonds Δ That's about like [asking if] we're ever going to have gender parity in our society. One can hope, that's all I would say. I think there are a lot of programs in place now that will certainly make, in the next generation, a significant impact. But we're going to have to wait a while to see that come to fruition. But, yes, I expect things to get closer to parity in the next decade. I do.

Dr. Kim Anne Kastens is a senior research scientist and adjunct full professor in marine geology at the Lamont-Doherty Earth Observatory of Columbia University, where she has been employed for 13 years. Among her professional interests are the structure and tectonics of oceanic transform faults, Mediterranean and circum-Mediterranean tectonics, processes of sediment transport in the deep sea, and innovation and nontraditional earth science education. She has lectured and consulted widely. Dr. Kastens has firsthand knowledge of alternative science career tracks within academia. In our conversation she describes Columbia's long-established and well-populated research track for PhD-level nonfaculty as a viable alternative to tenure track science.

Morse Δ Are you currently in a research track position?

Kastens Δ Well, it's very complicated. Lamont is a very complicated place. I have three simultaneous positions. One is a senior research scientist, which is the highest level in the Columbia research career track. Simultaneous with that, I have an adjunct full professorship in the department of geological sciences, which is different from adjunct professorships in most departments, in the sense that half of the people in the department are adjuncts, and those adjuncts are all Lamont researchers. The goodly share of people at the senior levels in the research career track have simultaneous appointments in the faculty in the department of geological sciences. We have full voting rights in the department and so forth. And simultaneously, I have a third position, which is called program leader, which is a quasi-administrative position, sort of providing leadership to the marine geology and geophysics group of researchers here. There's nothing straightforward about what I am.

Morse Δ Is it common that one person would be trisected?

Kastens Δ It's very common at Lamont, but Lamont is like no place else in the country. Nothing at Lamont is typical. It's a research piece of Columbia University, but it's on a separate campus. It's not part of the main university. And the research, our career appointments are the common appointment here. Less than 20% of the PhD-level scientists here are in the faculty "officers of instruction" job category. It's kind of unusual in the sense that in most university departments that do have research tracks, the research track people are in the minority, and here it's different. Here, we are the majority.

Morse Δ What precipitated the establishment of that track?

Kastens Δ I don't know. It is a Columbia career track, and I don't know how long it's been in existence. It's obviously a heavily used career track at Lamont and in the medical school, which are both on separate campuses. The main campus, the Morningside Heights campus, it's a more traditional situation, where in any given department there may be a few people in the research track and the majority of PhD-level people would be in the instructional track. I don't know how far back it goes . . . Lamont's been existence since the early 1950s.

Columbia University's career track for nonfaculty are called "officers of research" as contrasted with "officers of instruction." Within the officer of research career track, there are associate research scientists, research scientists, and senior research scientists. These positions are intended to parallel the assistant, associate, and full professor levels in the officers of instruction career path. All of these positions are senior to postdoctoral research scientists.

In general, the officer of research positions are not too bad, especially at my own institution, where it's the "normal" appointment held by the vast majority of PhD-level scientists. The salary ranges are similar to the faculty ranges (except you have to raise a higher percentage of your salary from grants and contracts). The positions have a great deal of freedom and very few bureaucratic restrictions, and especially appeal to a certain category of personality that doesn't like teaching and detests a highly structured working environment. It's also a good appointment for someone who does a lot of field-work that can't necessarily be scheduled to occur in the summer. Most of the benefits, such as retirement, medical plan, and other insurance, are the same as for other faculty.

Morse Δ Is there anything like tenure that's accorded to people in the research track?

Kastens Δ There is within Lamont. We have a thing called the senior staff, which is the research scientists, plus the senior research scientists, plus those faculty members who have their offices and do their research on our campus, who are within the department of geological sciences, who are tenured. For people in the research track, the process of getting on to the senior staff is very analogous to a tenure-review process. You have to be nominated from within the observatory, there have to be a bunch of supportive seconding letters, and then they ask for 20 external letters from all over the world. There's an ad hoc committee that meets with both inside and outside members that reviews your publications and your other contributions, and the letters from outside, and the letters from inside, and talks about what your future potential is, and so forth. Then it goes to a vote of the full senior staff, and from there it goes to a vote of the administrative committee, which is an oversight body which is a Columbia body. It's a process very much like a tenure review, after which you don't really have job security in the sense that you do when you have tenure, but you do have institutional backstopping of your salary. You're given, basically, salary in a bank account that you can draw on as you need it. And as you raise salary from external funds, additional money is put into your so-called bank account, against which you can draw if you want to do something new and different or if you run out of salary from grants and contracts. That's not Columbia-wide, just within Lamont.

Morse Δ Do you think other institutions are going try this setup?

Kastens Δ It's not unique to Columbia; there are quite a few universities that have research tracks for scientists. I don't really have any sense as to whether it's becoming more common or not. My guess is that it is, because, you know, jobs are so scarce for scientists right now. There are so few faculty jobs, and so many new PhD-level scientists are being cranked out, and there are so many people who are in postdocs right now, who are in their second or third postdoc, and there's no obvious place for them to go next. The difference here is that our postdocs are, if they want to be, on the first step of this career track that goes to these other job titles. My guess is that these other institutions are clogged up with postdocs, and the postdocs are doing really good work and making contributions to the university and to the department, and they're not finding starting, entry-level faculty jobs. Those universities will see a position analogous to our associate research scientist as a good way to keep those people in the department and doing science, or in the university and doing science.

My guess is there might be a trend toward establishing more of these research positions. I don't think it's a very good trend. Certainly in the earth sciences, there's already too many people competing for the research money that there is. I don't think it's good for the country, or for the universities, or especially good for the individuals, to establish a lot more people who are competing for that soft money funding. The other reason why universities might be intending to start such a track is that soft money people bring in a lot of overhead, because to do the same amount of science, the bottom line of the grant is inflated by the amount of the salary, and so the indirect cost line of the budget goes up in proportion to that. So, from a university administrator's perspective, probably the soft money people in the research track look like a financial good deal. That could be another reason why universities are starting such tracks. But I don't have any idea what the numbers are.

Morse Δ You mentioned some problems that women might have with this situation, especially with losing benefits during maternity leave.

Kastens Δ If an officer of research takes a maternity leave, she is paid her full salary for the period of time her doctor or midwife says she is unable to work. However, the money is taken out of the salary line of her grant or contract. So, in effect, she is responsible to her funding agency program manager for work accomplished during the time she was on maternity leave.

An officer of research may petition to work half-time to spend additional time with a young child after her maternity leave. She is paid half her salary while on that status. However, she loses all of her fringe benefits, including medical insurance. This is in spite of the fact that her grant or contract is being charged for fringe benefits proportional to her salary. In contrast, an untenured professor may petition for a career appointment for parents, in which status she (or he) collects half salary but retains all benefits. Alternatively, a professor at any level may apply for parental workload reduction, in which status they are paid full salary, retain full benefits, but work a reduced workload (usually relieved of teaching [but] continue to perform research).

Morse Δ What could be done within this kind of set up to make it more family-friendly? Could they be made parallel to the benefits accorded the instruction track?

Kastens Δ I don't think it's realistic, at least within this university, to exactly parallel them. The situation is quite different as far as the funding goes. One step that our institution has taken is that our promotion steps have time triggers on them. In other words, you can be a postdoctoral research scientist for 3 years and then you have to be considered for promotion to associate research scientist. It can be up to 7 years before you're considered for promotion to senior staff. Both of those have these time triggers attached to them. Lamont now has a stop-the-clock policy for those promotion triggers. If you take a maternity leave or, in fact, any kind of disability leave, you are granted extra time before you have to come up for promotion. However, you can come up for promotion on the normal time schedule if you prefer.

I think a very important step that could be taken that doesn't cost new money has to do with people who choose to work half-time. People who want to work half-time are still bringing in or getting paid half their salary, which is coming from their grant or contract. The way these things work is, say you take out $1000 worth of salary for a person, and pay that person $1000 worth of salary. Automatically 33%, $330 dollars are taken out of that same budget and put into the fringe benefits account. So the fringe benefits money exactly tracks the salary money from the perspective of the budget of the grant. However, part-time people don't get any of that money. Part-time people don't get any benefits, the money just goes into the fringe benefits pool and disappears. That money actually exists within the university. The National Science Foundation, or NASA, or whoever, has actually paid that money into the university, with the expectation that it was going to be spent on fringe benefits for the people whose salaries were being paid off that grant; but in fact the money's not being spent that way, it's just disappearing into the pool. I think it would be appropriate and fair and doable to pay prorated fringe benefits to people who are working half-time. My preference would be to see that done on a menu where you would get your choice, where you can add together the fringe benefits that you like in such a way that they don't exceed a certain dollar amount according to a formula. I can imagine that one person might choose to put their money into getting full medical coverage and say, "All right, I'll do without the retirement plan, I'll do without whatever." Or, you could split it up differently. You could say that the university would pay 50% of the costs of a medical plan and the individual would pay the other half. There are various ways you could do it, but the important aspect is that the money that's allocated for fringe benefits for that person who's working half-time ends up in her pocket, or potentially his pocket, rather than going into the pool.

Morse Δ Can people do good science on a half-time schedule?

Kastens Δ I know of a couple of people who have done it for a few years. I don't know if you could do it half-time for a whole career. I'm pretty convinced that you could do it half-time for a couple of years and then come back to doing it full-time, which is what would be appropriate for people in the research track here. This is a really high-powered institution, and so I think that it'd be tough to do science at that level, half-time, for a whole career. On the other hand, I'm not convinced that it's impossible, because if you think about how much time faculty members actually spend "doing science," they don't spend that much of their time doing science. They spend a lot of their time on committees and in a kind of teaching that you don't learn a lot of new stuff from—teaching courses that they've taught before and at a level that's below their threshold of new knowledge. The truth is that many people in faculty positions at other than the most flush research institutions spend far less than half their time doing research anyhow. So if you take that as a model, and your half-time person is careful to avoid those other things like committees and so forth, I think it might be possible. The only thing is, I can't think of anyone who's done it, so I can't say conclusively that it is possible. I don't think it's really been tried, but I think it's plausible.

Getting women into the sciences, with no jobs on the other end, is sort of a "bait and switch."

Dr. Bonnie Shulman teaches mathematics at Bates College, a small, well-respected liberal arts college in Lewiston, Maine. She has followed a nontraditional career path, leading a rather full life before she even entered college at the ripe old age of 30. Dr. Shulman is deeply involved with mathematics and science education reform and is a founding member of the Calculus Consortium, a group of mathematics educators working together on calculus reform.

Morse Δ How did you get into science? Before you went to college, you were a poet!

Shulman Δ Well, I was always good in math at school. I actually went to Bronx Science—a special high school in New York that you need to take a test to get into. Bronx Science had historically been a boy's school, but it was co-ed by then. This was during the 1960s. When I graduated from high school, I did not go on to college. At that time it was a big scandal, a big loss. Meanwhile, I traveled around the country, I was a hippie, I did a lot of that stuff. I got pregnant and had my daughter as a single mom. I ended up in Colorado. My other love had been poetry and writing. I studied at the Naropa Institute, with Allen Ginsberg, right the first year it started. It was tremendously exciting. I made a life for myself there, and I was on welfare, and I typed and transcribed a lot of tapes for a lot of poets, and we had little salons of writers and jazz musicians. I had a little printing press and published magazines and published some of my own poetry. Then, around the time I turned 30, well, for one thing, everybody had been saying to me, "You really ought to go to school." The only thing that anybody ever considered when they said things like that was English, because I was, you know, doing poetry and writing, and doing workshops in prisons. They thought that if I had real credentials, then I could make real money. That just sounded kind of boring to me, and I also had this attitude that I already knew it—I was already doing it—what would school have to teach me? I had a friend who lived in a teepee out in the woods (I had lived there with him) and he went back to school in geology. I saw this really change his life in positive ways. He was taking a calculus course as a prerequisite, and one night, by the kerosene lantern, I was looking over his shoulder at what he was doing. It was some calculus, and I said, "Oh, no, no, no. It goes like this." And he looked at me, and said, "How do you know?" And I said, "Oh, I must have taken calculus in high school." He was the first person that I had ever really told about this closet background that I had. At one point, when I turned 30, I was going through all this angst about, "Oh, what am I going to do with my life?" and about whether we were going to have to buy used clothes forever. I was typing all these papers for these students, and 4 years later they're making five times as much money as me, and I'm basically writing these papers without ever taking the courses, and [my friend said], "Well, you should go back to school in science! It's the only thing that makes sense to go back to school in." And for some reason, that really took in my mind at that point. A month later, I was enrolled in school and ready to go. It was like a hurricane.

My daughter was 9 years old, and I had raised her alone, and had done all these wild things, so school did not intimidate me. I figured I could do anything. I had all of this confidence, which is a really good thing, because I had no idea how hard what I was taking on was. I decided to go to school and get my PhD. Starting as a freshman, now, at 30 years old, I said, "All right, I'll have a 10-year plan." I was looking in the catalog, and going down, you know, "Anthropology, A. No, no, no." There was one other, maybe archaeology. The next one was astrogeophysics. I thought, hm, that looks interesting, and there were all these cool courses in cosmology. So I thought, "OK, I'll major in that." And that's how I started school. I was still good in math, as it turns out. Although I took all the astronomy and undergraduate courses they had and was very interested in that, I was advised that I should really be in mathematics, that it was really my strong suit. I think it was probably pretty good advice, although the woman who advised me was a professor of mine, who taught the most upper-division astrophysics course I had taken, and I turned around and became her graduate student. After I passed my comps in mathematics, I worked with her. I got a degree in mathematical physics, and she was my advisor. Sometimes I think she wanted a mathematician on board, so she sent me off to get trained, and then I came back and worked for her! That's the short version of how I got started.

Morse Δ What do you think about the glut of new PhDs?

Shulman Δ I'm very lucky, I'm right in the middle of that. My graduating class, and the two before it, and the ones coming after are, you know, definitely part of that. But still, if you look at the percentages, 85% of the people get jobs. The problem is that most of them are not getting the jobs they want. They are not getting into the research institutions they wanted to be in, and the grants—the money's much harder to find. Then there are many more that aren't getting jobs. There is a sense of entitlement, I think, which I don't necessarily go along with. But people feel like promises made to them have been broken. I think in some ways, they have a right to feel that way. They've been lied to. I don't think anyone is necessarily entitled to a job. There are some classist issues there that I have some problems with.

I was both lucky, and in some ways, the fact that I was, as they say, a nontraditional student and had experience out in the work world worked in my favor, because I didn't fool around. I knew how to write a resume; I knew the kinds of things you do if you really want something in order to go after it. I think there's a lot of bad or total lack of advising of younger people. They flounder around. Many of them send out a 100 copies of a form letter [to prospective employers]. I very carefully chose the kinds of institutions I wanted to work at; I also chose not to go on the fast track in high-profile research. I really wanted to teach. So I went to a primarily teaching institution that also values research, but it's about a 50–50 split. There are places where it's even less than that, and you don't get to do any research at all. So I researched and picked the kind of places that I wanted to go to. And then I wrote, along with my resume, very carefully constructed cover letters to each school, to let them know that I really knew who they were and that I really wanted to be there. I tried to stress my unique qualifications and turn them into strengths. There could have been discrimination because of age and because of my nontraditional background. In some ways, being a woman really is an advantage, because of all the affirmative action. But that's a mixed bag too. It worked in my favor. I know that the department that hired me had basically been told, "Thou shalt hire a woman." I had a lot of interviews, but this was the only offer I had. It turned out to be a really good match. And that's the luck part. So, there are jobs, but it is a tough market out there. I feel for a lot of my friends.

Morse Δ Do you think that the country is losing out on an opportunity? Maybe these people feel some inappropriate entitlement, but we do have some very bright individuals, who are highly trained and are either unemployed or underemployed. Are we throwing away an opportunity to use their skills?

Shulman Δ Yes, but my ideas about how to use them might be different than the government's! There's a famous line from an Allen Ginsberg poem, "I saw the best minds of my generation . . ." It goes on to talk about running stark naked through the streets. I feel that, while we do train some of the best minds of our generation maybe a bit too narrowly, we nevertheless train them; yet we have all these problems that need to be solved. Who else but the best minds of our generation should think about those problems and come up with solutions? First of all, we aren't really training them to do that or providing them with opportunities to do that. Some arguments could be made that we're buying them off so that they *don't* think about problems. The solutions that they come up with might be somewhat threatening to the status quo.

A lot of physicists are ending up on Wall Street, for instance. They have certain kinds of skills that are enabling them to make lots of money there and they are in high demand. If what you believe is that when you graduate, you'll be getting a job and making money, then maybe this would be attractive. You may not get to do your physics, but you're getting some kind of life.

We're just kind of casting them [new science grads] adrift, especially with the values that were instilled along with this education, and there's very little hope that many of them are going to end up spending much time either doing science, which at least, perhaps, might have some useful applications, or, you know, just thinking in holistic ways about how to deal with the mess we've got. I think it's a tremendous loss. That's why I'm so committed to education. I want to mold young minds, or provide opportunities for young people. When you're young is when you're idealistic and fired up. If nobody tells you how it's going to be and what you can do with that, then it turns to cynicism and you don't do anything with it. So I try to work with the young people I meet and send them off to do good things.

Morse Δ Do you think women differ from men in any sense in that way? Are women starting their careers as even bigger idealists than men?

Shulman Δ From the student perspective? I'm reluctant to say yes or no to that, because I think that the ideals that they have might differ, but I think being idealistic is just from being young. What those ideals are might be gender linked, but I don't feel the energy level being any less in men than in women. The dragons they want to go out and kill may be different, or just that metaphor itself—they may just want to go kill dragons rather than save the world, I'm not sure.

Morse Δ Do you have any feeling that a science education might actually drum the idealism out of young people?

Shulman Δ Oh, yes. I think that you're acculturated. In some sense, that's the most important part of your education. I think we, talking now as someone who went through this process and who was educated that way, are trained maybe *not* to think. Social responsibility and scientists, for instance: There is a lot still embedded in the culture that says, "Hey—you're after a pure pursuit of knowledge. Let somebody else worry about what happens with it. That's not your job. In fact, that's why there are people trained in political science. Your job is just pure pursuit of knowledge, and science is value-free, and that's what's so cool about it." There's a delinking of values and science, which I think is first of all impossible and not true, and leads to very dangerous kinds of knowledge. Knowledge without wisdom is a very dangerous thing.

Morse Δ This is a feminist critique.

Shulman Δ Yes.

Morse Δ With this in mind, if we had more feminist scientists, men and women, what would science begin to look like?

Shulman Δ That's a fun question. Of course we have to define what we mean by "feminist," "science." Two different categories. And then what we mean by linking them. Some people may say that it's an inappropriate mixing of categories even to begin with. I'll say a few words about what I mean by that and then I'll spin off on some vision things.

A radical critique of empiricism is an important part of what is happening in feminist critiques. It's one of the more interesting aspects of it for me, in terms of answering this question. One vision I have is of a science unashamedly linked with values, where we make conscious and explicit what values are there. Science is done with the goal of intervening. It's not just pure pursuit of knowledge: we intervene. We use our knowledge to intervene in the ways of the world. So let's think of how we want to intervene, and let's be very up front about that. So any scientific work includes a kind of a listing of here are the biases, or here are the values, or here's explicitly what we're trying to do. And then we might come up with different kinds of knowledge, as we have different kinds of intervening strategies. I am not at all proposing that we throw out the notion of empiricism or saying that we can't have reliable kinds of knowledge about the world. We have one very narrow window on what we mean by "reliable." Let's open it up. There could be other ways, other unexplored ways of getting reliable knowledge about the world and using it. There is not pure pursuit of knowledge, so let's say what it is we want to do and why, and open up the possibilities of what we want to do. Then I think that we can use this power, because that's what it is, more consciously and more to our betterment.

Morse Δ What I perceive to be happening now is that there is a denial and an abrogation of responsibility for things that are happening with scientific knowledge. So, if we consciously admitted to it . . .

Shulman Δ Make it part of science. Include it as part of what we mean by science, yes.

Morse Δ Can women do that?

Shulman Δ We can all do that. What do women, perhaps, and other previously excluded groups have to offer? You find this in feminist standpoint theory. By virtue of having been excluded, they have a different standpoint, and so they bring new perspectives. That's something important that excluded groups have, but I don't think that women have a corner on the market. For instance, there was my office mate, who had a very different perspective on things. He was a man, but he was an unusual man. In some ways we are doing it, but it's individual and piecemeal. We need to organize. You sense reluctance in me. You know, it's Margaret Thatcher.

Morse Δ Her name pops up over and over again!

Shulman Δ Being born a biological woman doesn't necessarily mean anything. That's not the point. There's something about being part of an excluded group that can lead to new perspectives, but doesn't always.

Morse Δ There are also levels of exclusion.

Shulman Δ If you are completely excluded, then you have no power. It can also lead to bitterness, cynicism, and destructive kinds of energy as well. But all of that having been said, I do find that by and large it's women that are wanting to do this kind of work, at least right now.

My last interview was with Dr. Mindy Kurzer, an assistant professor of nutrition in the Department of Food Science and Nutrition at the University of Minnesota. She received her PhD degree from the University of California, Berkeley, in 1984, and spent the following year in Europe as a NATO postdoctoral fellow. While in Europe, she worked at the University of Odense, Denmark, and at the National Institute of Nutrition in Rome, Italy. The dietary regulation of estrogen metabolism is Dr. Kurzer's primary research area. During 1992–93, she received an International Life Sciences Institute-Nutrition Foundation Future Leader award for her work, and in 1994 she received the University of Minnesota College of Agriculture Distinguished Teaching Award. She has spoken on radio and appeared on the PBS television program "Newton's Apple." She first came to my attention as a panelist at a university-sponsored forum on feminist science.

Morse Δ You mentioned that you always perform an experimental procedure on yourself before you begin clinical trials with others. Is that because you are a woman? Or is that because you are Mindy?

Kurzer Δ I always view it because I am Mindy. I don't know of many other people who do that, male or female. I got that from my mentor, my faculty advisor for my PhD. She was a woman, and she did the same thing.

Morse Δ Are there feminist male scientists out there who would be doing science differently if they felt they could break the mold and not lose tenure or be scorned by their colleagues?

Kurzer Δ Yes, in general, I definitely agree with that. When it comes to issues related to sexism and feminism, it's not strictly a gender division. There are certainly men who are more feminist than some women that I know, so it definitely crosses. And when it comes to science, I do think it's true. The problem is that science, as it is now, does not attract people who come from an alternative perspective—men or women. Because of that, I think there aren't a whole lot of people who would be interested in joining this dialogue. Of the people who there are, I'm sure that the vast majority if not all of them are women, because science doesn't attract people with an alternate perspective. Those with alternate perspectives are probably women who tend to lean more toward a feminist perspective because of our gender.

Morse Δ How come science doesn't attract people from alternate perspectives?

Kurzer Δ Because the system is very rigid and very narrow. It's extremely competitive and so the kind of people who go into academic science must be very competitive people who are willing to abide by the rules. For those of us in academia, the rules are very strict, and people who don't follow them can wind up in trouble. Academic freedom, and the tenure system which was developed to protect academic freedom, allows academic people to hold alternate intellectual perspectives. But when it comes to alternative social perspectives that really challenge the system, those are not really nurtured very much in academia. Do you know what I mean? There's a slight difference, I think, between academic or intellectual freedom and the freedom to really challenge the very system that we're within.

Morse Δ So as an insider . . .

Kurzer Δ It's difficult, until you have tenure. Then I think people who make it through the system—which involves a lot of compromises and a lot of accepting the framework—once people get tenure, then they're in more of a position to challenge the system. But by then, people tend to be older and have gone through a lot. Things may have shifted a little bit, people sometimes tend to move a little bit more toward the mainstream, the dominant culture, as they age. It's less likely that challenges would arise at that point.

Morse Δ What brought you into science in the first place? You seem to come from an "alternate perspective."

Kurzer Δ I'm secretly a social scientist—I'm not really a life scientist. I studied history and philosophy when I was in school, and when I was an undergraduate I took no science classes at all. I graduated from college in 1972, and so I went to school in the late 1960s and early 1970s, when there were no breadth requirements at all. I was planning to go to law school; that was really my direction. I was good in science when I was younger and I enjoyed it, but I had to make the choice, so I made the choice to go more in a political, social kind of direction. After college I applied for law school. I was about to go, and realized that I didn't really want to go because my interest in law was from an idealistic, social change perspective, and I realized that that isn't what the law is about. The law is about manipulation and finding loopholes and things like that, and I realized it wasn't for me. So I took a few years off and decided to take a couple of science classes at night just for fun to see what I thought about it, and really enjoyed it—I loved it. And I decided that that was what I wanted to do. That explains some of my more political activist approaches to some of this, because that's really my background from when I was young.

Morse Δ And you've seen the world, too.

Kurzer Δ That's right. I've also lived abroad a couple of times. After my PhD, I did research in Europe for a year, and I've been there a couple of times since. And that has broadened my perspective out to an international perspective, even beyond the kind of typical American view.

Morse Δ I've been told that European scientists have a more livable schedule and still do good science. Did you find that to be true?

Kurzer Δ Yes. I was in Scandinavia, the longest period of time I spent was in Denmark, and I certainly found that to be true. In the department that I was in, people took coffee breaks twice a day—the whole department got together —and it was expected. I couldn't do it, because I was coming from a very different background, and I felt like I didn't have time and I'd rather get home early than sit a half an hour and talk to people. But that was clearly what everybody did. They went home early, and everybody took their vacations and had other lives, and it did strike me as very different. However, the European science system is moving more and more in the direction of the US system.

Morse Δ Why?

Kurzer Δ I think it's partly because of tightening funds. In many other countries, there are federal government medical research funds that fund scientists, and there aren't as many scientists as there are in this country. We have a much bigger country, we have a huge number of higher-education institutions and government labs. The old system in many countries in Europe was set up in such a way that anyone who made it to the position of a scientist was pretty much funded and they were guaranteed of it. But I think now, there's less money. They're becoming more competitive and they're actually having to compete for grants. And I actually argued with the man who I was working with at the time, because he thought that was a very good thing. He felt that the competition resulted in better-quality science, and that when people assume that they're going to be funded, they become lackadaisical, they slow down, they don't work as hard. We argued about that, because I told him that not everyone works well under that kind of pressure. He really believed that increased competition was a good direction that was going to improve their science.

Morse Δ Were there many women scientists there?

Kurzer Δ No. Very few.

Morse Δ How do you think a more reasonable workday might affect women's entry into the sciences?

Kurzer Δ I think it would make a huge difference. Because the culture in academia is—how can I put this—the ideal role model of the successful scientist academic is someone who's a workaholic, who spends all of their time here. That's *very* admired. People who come in on weekends, people who work evenings—there's tremendous admiration and respect for that from most people and a lot of accolades go to that kind of person.

Morse Δ From women too?

Kurzer Δ From the women who identify enough with the system so that they are really part of it, yes. Many, many women here identify with the system, have learned how to succeed within it, have learned how to adapt to it, have given up tremendous sacrifices, much more than some of the men, and yet they still identify with the system. They still want to be successful within the parameters that are set up, and so they would also admire people who work 50 to 80-hour weeks. That's expected. It's not acceptable here to say no to anything. And so there's a real push to do everything, and to kind of over-achieve, and basically to have no life.

I think that a lot of young women don't even think about academia. And if they do, and they get PhDs, they look and they see what's involved, they may wind up getting married during graduate school and either having a family or getting into a relationship which has responsibilities. They may buy a house and feel tied to the area, and they aren't as inclined to put their career first as men are in that same situation. And because of the demands of the profession that really, I think, ought to be modified, because of the demands, that they really start thinking about other things. So if the profession changed enough that it allowed people to succeed and have other aspects of their lives developed as well, I think there would be more women.

Morse Δ Who is going to make those changes?

Kurzer Δ Well, I don't know. I think some of the changes are occurring already, very, very slowly. But the changes are certainly occurring already, because my generation of faculty people, I think, are less willing to accept the older model than the older people were. I think that even people who are not necessarily feminists per se have other things in their lives that they want to develop. Men and women both don't want to give up the time with their families as much. So I think that the culture is shifting a little bit, and some things are happening. For example, at the U of M there's a policy that's been instituted very recently that will allow a tenure-track faculty person to stop the tenure clock for a year if they have a child. And that's a man or a woman, and it could be a birth or an adoption. And that will allow them an extra year toward tenure. Usually, you know, it's a 7-year clock, and when it ends you either get tenure or you're gone. So to have an extra year is a very significant thing. And that's a really important step, I think, to acknowledge the fact that we need to have a system here that allows for people to have children and still be able to succeed, and not to have to make a choice to be a neglectful parent or to fail in their job. That kind of a compromise really helps people be able to do both. It will have a small effect, but I think it's really significant that that even happened.

Morse Δ Some people say that women are more moral than men as scientists.

Kurzer Δ My personal opinion about the whole issue of the innate differences between men and women is that I believe that most of the differences that we ascribe to men and women are culturally defined. Women do tend, and this is a big generalization, but because of the difference in social conditioning between men and women, women do tend to be more concerned about other people. Women tend to be more concerned about relationships, they tend to be more concerned about the impact of their actions; whereas men tend to be more self-centered, they tend to assume that everything they do is fine and not to question it. Women tend to question themselves too much. Men tend to question themselves not enough. I think that those are the kinds of things that might result in someone saying that women are more moral. I don't believe that there's an innate difference.

Morse Δ What about the future? What's going to happen? Twenty, 25, 50 years from now, will there be sexual parity in the sciences?

Kurzer Δ Well, I think that there are going to be some changes. We will be further along the path toward parity or toward the kind of changes that we've talked about. I don't know that we'll feel as though we're totally there. But I really do believe that we'll be further along on the path. Even things like, and this isn't strictly feminist, but I do believe that it's a very related issue: the National Institute of Health has this past year opened up an Institute of

Alternative Medicine. And they've actually allotted money, it's a very small amount of money, for the purpose of studying alternative health treatments, like acupuncture, macrobiotic diets, and homeopathy. That's a real shift toward them acknowledging, OK, we're not just going to ignore this stuff, we're going to try to study it at least. Just that little bit of opening up to some alternative way to view medicine, I think, is another kind of step toward opening up something different.

Morse △ You mentioned your field was about half women, and that none of them was, as you termed it, an "ally."

Kurzer △ In my field, I would say that there are very few women that I know who would identify as feminists and really be concerned. Even if they identify as feminists I don't think these issues are priorities to them. Most women in science tend to really identify with the system and accept it.

Morse △ So they throw their kids in day care and do their jobs.

Kurzer △ A lot of them, yes.

Morse △ And they don't regret it.

Kurzer △ Well, I think many of them feel conflicted and angry but they accept it. They don't question it enough to try to take some action to change it.

8

Women Changing Science

What is to be done with Western science? Shall we leave it alone, or can we honor the experiences of the women whose words appeared in this book, along with those of their sympathetic male counterparts, and agree that changes need to be made? In this last chapter I will summarize some of the suggestions that my respondents advanced as possible solutions to the dilemmas that science now presents and will offer up some of my own.

It is essential to acknowledge that the bulk of this book dealt with the scientific status quo in the United States, and that the solutions being put forth may already be in place in other countries. A fascinating resource for considering the science climate for women around the world is a special report published in 1994 in *Science*,[1] which addressed science culture not only in the United States, but in Italy, Portugal, Great Britain, Turkey, Sweden, India, and the Philippines. As we look to the future there is no point in reinventing the wheel—women scientists throughout the world would benefit from an open dialogue across cultures and among nations.

Science Education

Kindergarten through Twelve

Girls, and smart kids generally, are not getting the full benefit of their formative educational years. Teachers, school administrators, and parents must evaluate not only the curricula being presented to children, but the style in which it is presented. Of overarching importance in all these suggestions is the elimination of sexism, sexual harassment, and racism in the classroom.

There are several specific steps that schools can take to support girls' interests in math and science, proposed by the American Association of University Women and summarized in Chapter 2. In the AAUW's nationwide poll, *Shortchanging Girls, Shortchanging America,*[2] the authors state that,

> Schools transmit gender bias in the thousand and one signals they send girls and boys about what's expected of them. These expectations determine how girls and boys are treated, how they're taught, and ultimately how they're tracked onto different paths through their schooling and into their careers. In dozens of separate studies, researchers have found that girls receive less attention, less praise, less effective feedback, and less detailed instruction from teachers than do boys.

Solutions to these problems are stated within the problems themselves. Girls need more attention, more praise, instructive feedback, and detailed instruction from their teachers. The AAUW has created publications and other programs to assist schools and educators in addressing the basic gender inequities found in their classrooms.

In addition to providing better instruction, schools need to embody a spirit of academic acceptance. Girls who manage to be academic achievers must not be ridiculed by their peers or ignored by their communities. Teachers should be vigilant in stopping taunting or other humiliating behaviors toward achievers of both sexes. The emphasis placed on athletic achievement should be extended to academics. Newspapers should highlight the work of intellectual high performers, just as they do local sports stars. Schools should ensure budget equity for extracurricular academic activities, such as debate or biology club, by bringing their budgets in line with sports budgets and encouraging students to become involved in these activities by making them rewarding, interesting, and accessible. Parents should form support networks for organized academic activities, just as they have done as sports boosters, band boosters, and so forth. Children who wish to participate in academic summer camps should be allowed to do so, supported financially and programmatically by schools and parents to the same extent as children who attend sports camps.

In junior high and high school, counseling programs must provide honest and accurate information to students, not only on what they must achieve in order to enter the college of their choice, but on the most current prospects for various career choices. Many "counseling" offices appear to be swamped by the needs of students with social and behavior problems, making it impossible for them to spend time on career development. Author Peggy Orenstein[3] describes the caseload of one school counselor in her book, *SchoolGirls: Young Women, Self-esteem and the Confidence Gap*. The counselor, charged with following more than 300 students each year, reported that her caseload overwhelmed her, and that she had no time "to do much more than discipline" the students. If a counselor, especially in a school where students come with myriad social and family problems, does manage to free up some time to spend with career development, it is often spent administering written aptitude tests, poor substitutes for in-depth exploration of student's career interests and abilities. Perhaps most important is the idea that students deserve to understand that their performance and coursework as early as junior high can directly affect their college admissions prospects, scholarship eligibility, and intellectual preparedness for undergraduate work.

In junior and senior high, as well as in undergraduate education, the concept of single-sex schools—or even single-sex classes within gender-integrated student bodies—merits further attention. As discussed in Chapter 2, girls who attend all-girls schools show significantly higher aptitude and interest in science and mathematics than girls who study these subjects alongside boys. I view the value of gender-segregated education as an interim solution to the long-term problem of gender inequities in school: single-sex schools, even though they now appear to provide better educations for girls, are an artificial environment where neither sex gets to interact within a realistic social model. Given, however, the chaotic milieu in many US schools, where violence, sexual harassment, and precocious sexual activity among students are wreaking havoc on learning, the institution of single-sex conditions may be, at least for now, the best way to ensure a civilized and attentive learning environment for girls.

Undergraduate Years

Undergraduate science departments should institute mentoring programs, informal and formal scientific support networks, and career seminars for women involved in science courses or interested in science careers. Academic counseling ought to be honest and personalized, as described for K–12 education. Universities should continue to recruit women as professors and teaching assistants and to provide role models to women students. Men who wish to mentor female students should be encouraged to do so. Blatant sexism in the classroom (such as the resistor code mnemonic described in Chapter 3) must be eliminated. Teaching styles should emphasize inclusion rather than exclusion: Science professors who seem intent on "weeding out" low performers rather than providing a rewarding education to each student in their class might be given poor evaluations and appropriate retraining.

A point of some importance can be made about undergraduate science training: Students who graduate with strong backgrounds in science would be well-served, again, by thoughtful and well-informed career counselors, who would keep abreast of job developments for students in these fields and could offer accurate appraisals of the career market. If math majors would benefit from further training in business, or if environmental studies majors might profit from a law degree, then these options should be presented at least as compellingly as the traditional counsel that a PhD is the only suitable graduate field for a science major.

Graduate School

The most important suggestion to improve science vis-à-vis graduate school is truth in advertising: end the deceptive practice of training graduate students for "traditional" PhD-level employment. Universities should formally acknowledge that future careers will make use of MS and BS degree holders, and that academic careers will be the province of relatively few scientists. Graduate institutions should actively present career alternatives to students pursuing advanced degrees. It is not enough anymore, according to Ohio State University's Zachary Levine,[4] to say to a student who has just defended a dissertation, "Physics gives you a great general education. Good bye and good luck." If universities are going to reap the mammoth benefits of the scientific research machine, they owe it to graduate students to build job preparation into their programs.

The Scientific Workplace

The Academic Workplace

PhD-level employees, women and men, could benefit from a number of changes in the academy. First, since so many new PhDs are employed as postdoctoral fellows, and since this career step is bound to lengthen as science funding scarcity grows, institutions should consider changing postdoctoral status to a semipermanent position. Postdocs, especially after completion of one or two appointments, should be granted the status of mid-level scientist with all of the benefits accorded tenure-track staff. Of course, funding would still be contingent on research grants and appointments would still be on contract, but postdoc-level scientists should not be forced to work on a strict term basis, relocating with their families from institution to institution every 2 or 3 years. The Columbia University model, described by Kim Kastens in Chapter 7, would be a useful guide.

In addition to stabilizing the nature of a postdoc, we should consider leveling the playing field for mid-level scientists who propose important research projects. An initial recommendation is that federal science funders recognize the "hidden" scientists who have meritorious ideas, but who may be stuck in the role of postdoc or even technician in a principal investigator's lab, and are therefore disqualified from applying for research funds. An increase in available funding, plus a loosening of the regulations governing who is able to apply, should benefit not only the large pool of underemployed, PhD-level scientists, but the overall quality and quantity of research output.

Professors, both on and off the tenure track, could benefit from a science workplace that shook off the trappings of 1957: Scientists should not be expected to put in 50 to 80-hour weeks. Women, especially, have taken on the responsibility of caring for children, families, households, and communities; to expect women to do all of this while essentially living at the lab is insane. There is no reason that science cannot be performed quite well on a less demanding schedule. Institutions should look to France, Sweden, Mexico, and many other countries to see how more cooperation, more flexible scheduling, and more institutional support for worker's personal lives and communities can benefit the science being produced there. Accusations of poor-quality science coming out of low-pressure laboratories can (and should) be shown as inconsistent with reality. US labs should experiment with more sustainable working models to see just how scientists' productivity would be affected.

Universities should reconsider the role of tenure. If it is determined that tenure is valuable, then the tenure schedule should be revised to accommodate the realities of today's families. If the tenure clock is ticking during the all-important and all-consuming childbearing years, women and men scientists who become parents during this time are vulnerable to failing either in their parental roles or their academic ones. Stay-at-home wives do not exist for most scientists—and even if having one parent staying at home with the children were a desire of the couple, they would be very hard-pressed to afford it on an average entry-level science salary. Again, institutions should experiment with part-time professional positions, flex-time, and other creative methods to support inclusion of a diverse scientific workforce.

Nonacademic Employment

Many of the ideas listed above would also apply to corporate science positions. In fact, given private corporations' interest in increased productivity through workplace innovations and their relatively quick ability to institute change (as compared to the cumbersome nature of academia), one might expect that science positions in the private sector might evolve to offer very livable and rewarding experiences. Scientists already receive preferential or at least different treatment in some corporations, where they can get away with less formal apparel and working habits than can their nonscientific colleagues. Yet scientists still suffer from the "hours worked equals level of commitment" difficulty. Experiments with more progressive hours policies, child-care options, job-sharing, and telecommuting are all welcome.

Workplace sexual harassment is still a difficulty in academic and non-academic workplaces, though in many cases it seems to be subtle. Employers, fearing lawsuits, have begun to institute sexual harassment training and work-force diversity seminars; however, these have limited efficacy where a top-down culture of respect does not exist. Recurrent sexual harassment discussions are essential in organizations where numbers of new individuals are entering the workforce on a regular basis (such as in universities), or where scientists of diverse cultural backgrounds are working together.

Science Culture

Here, the barriers to a gender-balanced and productive scientific workforce are erected by science funders and scientists themselves. Many of these problems were addressed throughout the book; here, we'll dissect a few.

The value of *competition* in scientific endeavor is tricky. One wants to encourage scientists (and everyone else, for that matter) to do their best, and competition is one way of providing incentive to do so. Yet competition obviously turns off many would-be scientists, and these individuals might be outstanding contributors to the field. Cutthroat competition in the classroom is unnecessary and should be replaced by alternate learning strategies, including group projects and individualized learning programs. At some point, of course, students will need to be guided into appropriate levels of advanced study; but this can be done based on individuals' aptitudes and performance, rather than through a narrow and inflexible gateway. Last, competition should be carefully monitored in the educational setting to ensure that students are not being negatively affected by it. Individuals should be encouraged to compete with themselves, rather than working against their classmates.

Competition should also be addressed in terms of its negative effects on working groups. Incentives to collaboration, in the form of grants to multiple scientific groups working on related problems, should be a priority. Funders might also consider bankrolling publications that would specialize in cross-disciplinary topics, making the possibility and benefits of collaboration obvious to readers. Since, as one respondent told me, professional journals are multiplying at a dizzying rate, these multidiscipline publications would hopefully be more interesting and appealing than average.

The *intimidation* problem resides in a couple of areas. On the interpersonal level, scientists should recognize that intimidation is childish. On the professional level, intimidation is counterproductive. Intimidation is really one of the most off-putting behaviors used by scientists, and it seems to be running rampant in the academy. I am not basing this assertion only on the reports I received from scientists. On surveys or in e-mail messages, I received from time to time some cutting or snide remarks from scientists who, after I followed up with them, were obviously sane and normally pleasant individuals. Knee-jerk arrogance and attacks on my motivations were completely uncalled for in these situations, and I could only imagine how difficult it would be to work in an environment where this type of behavior was the norm. It would appear that some scientists could benefit from instruction in subtlety and manners and from a reminder that basic respectfulness is a virtue. This would have to begin at very high levels, where egos can be quite inflated, and if successful would trickle down through the labs and into the classroom, where students are learning how to behave as scientists.

The intimidation that is intrinsic to scientific work, such as in response to a presentation of research, is a bit more difficult to change. In this case, conference sponsors could experiment with ground rules for presentations, perhaps limiting audience questions to written ones selected by a moderator, who could play referee and weed out the "questions" really meant to be openly inflammatory.

The Values and Goals of Scientific Inquiry

Last, I'll refer back to a subject that was discussed in Chapter 1. Does science serve the public in general, and women in particular? Do the questions that science seeks to answer have validity across cultures, income levels, races, and all of the other characteristics that define the human experience?

The feminist critique of science has opened the gate of inquiry that Thomas Kuhn[5] unlocked. Ecofeminists and health activists have shown quite clearly that women's needs have been neglected and that women's natures have been violated. Women and men in the environmental and peace movements are questioning whether science has really contributed to human progress. Third World peoples are calling attention to the injustices foisted on them by technological progress. These voices cannot be ignored, nor can they be trivialized.

Change is always difficult in an entrenched system such as Western science; but change is inevitable, for science does not function in a vacuum. Even if not one more woman entered into the study of science, women's progress in the other areas of public life would ensure that the values and end products of science would receive more careful scrutiny. Women are taking permanent places in government, in business, in advocacy roles, and in the decision making of their families. Through their votes, science-funding priorities will be altered. Through their boardroom decisions, research and development dollars shall be spent. Through women's activism, the goals and relevance of scientific inquiry will be scrutinized and publicized. And the decisions that women make on behalf of themselves and their families in the marketplace will define the value of technology. Of greatest importance, however, are the expectations and values that today's women bring to their careers in science and those of the many, many women who will follow.

Notes

Chapter 1

1. Sandra Harding, *Whose Science, Whose Knowledge?* (Ithaca, NY: Cornell University Press, 1991), p. 3.
2. Londa Schiebinger, *The Mind Has No Sex? Women in the Origins of Modern Science* (Cambridge, MA: Harvard University Press, 1989), p. 66.
3. Ruth Ginzberg, Uncovering gynocentric science, in: *Feminism and Science*, Nancy Tuana, ed. (Bloomington and Indianapolis: Indiana University Press, 1989), p. 71.
4. Cynthia Fuchs Epstein, Constraints on excellence: Structural and cultural barriers to the recognition and demonstration of achievement, in: *The Outer Circle*, Harriet Zuckerman, Jonathan Cole, and John Bruer, eds. (New York: W.W. Norton & Company, 1991), p. 249.
5. Hilary Rose, Beyond masculinist realities: A feminist epistemology for the sciences, in: *Feminist Approaches to Science*, Ruth Bleier, ed. (Elmsford, NY: Pergamon Press, 1986), p. 57.
6. Evelyn Fox Keller, The gender/science system: Or, is sex to gender as nature is to science? in: *Feminism and Science*, Nancy Tuana, ed. (Bloomington and Indianapolis: Indiana University Press, 1989), p. 41.
7. Ruth Hubbard, Science, facts, and feminism, in: *Feminism and Science*, Nancy Tuana, ed. (Bloomington and Indianapolis: Indiana University Press, 1989), pp. 119–131.

8. Jeffrey Goldstein, Nonlinear transformations: General comments on system processes, Proceedings, *Chaos Network Conference*, 1992, p. 117

9. Karen Lehrman, Off course, *Mother Jones*, Sept./Oct. 1993, pp. 45–51, 61, 66, 68.

10. Susan Faludi, *Backlash: The Undeclared War against American Women* (New York: Crown Publishers, 1991).

11. Katie Roiphe, *The Morning After: Sex, Fear, and Feminism on Campus* (Boston: Little, Brown & Co., 1993).

12. Christina Hoff Sommers, *Who Stole Feminism: How Women Have Betrayed Women* (New York: Simon & Schuster, 1994).

13. Faludi, p. 319.

14. Anne Eisenberg, Women and the discourse of science, *Scientific American*, July 1992, p. 122.

15. Ruth Hubbard, Science, facts, and feminism, in: *Feminism and Science*, Nancy Tuana, ed. (Bloomington and Indianapolis: Indiana University Press, 1989), p. 127.

16. Nancy G. Slack, American women botanists, in: *Uneasy Careers and Intimate Lives*, Pnina G. Abir-Am and Dorinda Outram, eds. (New Brunswick, NJ: Rutgers University Press, 1987), p. 95.

17. Schiebinger, *The Mind Has No Sex?*

18. Londa Schiebinger, The history and philosophy of women in science, in: *Sex and Scientific Inquiry*, Sandra Harding and Jean F. O'Barr, eds. (Chicago: The University of Chicago Press, 1987), pp. 7–34.

19. Carolyn Merchant, *The Death of Nature* (San Francisco: Harper & Row, 1980), p. 273.

20. Schiebinger, The history and philosophy, pp. 7–34.

21. Jessie Bernard, *Academic Women* (New York: Meridian Books, 1964).

22. Margaret Rossiter, *Women Scientists in America: Struggles and Strategies to 1940* (Baltimore, MD: Johns Hopkins University Press, 1982).

23. Carol Cohn, "Clean bombs" and clean language, in: *Women, Militarism, and War: Essays in History, Politics, and Social Theory*, Jean Bethke Elshtain and Sheila Tobias, eds. (Savage, MD: Rowman & Littlefield, 1990), p. 35.

24. Helen Longino, Can there be a feminist science? in: *Feminism and Science*, Nancy Tuana, ed. (Bloomington and Indianapolis: Indiana University Press, 1989), pp. 45–57.

25. Harding, p. 143.

26. Margie Profet, Menstruation as a defense against pathogens transported by sperm, *The Quarterly Review of Biology*, Vol. 68 (September 1993), pp. 335–86.

27. Harding, pp. 309–310.

28. Carol Gilligan, *In a Different Voice* (Cambridge, MA: Harvard University Press, 1982), p. 2.

29. Doreen Kimura, Sex differences in the brain, *Scientific American*, Sept. 1992, pp. 118–125.

30. Winifred Gallagher, How we become what we are, *The Atlantic Monthly*, Sept. 1994, p. 42.

31. Maria Mies and Vandana Shiva, *Ecofeminism* (Halifax, Nova Scotia: Fernwood Publications, 1993), p. 183.

32. Katha Pollitt, Are women morally superior to men? *The Nation*, Dec. 28, 1992, pp. 799–807.

33. Harding, p. 113.

34. Harding, pp. 54–55.

35. Noretta Koertge, Are feminists alienating women from the sciences? *The Chronicle of Higher Education*, Sept. 14, 1994, p. A80.

36. Steven Goldberg, Feminism against science, *National Review*, Nov. 18, 1991, pp. 30–33.

37. Merchant, pp. xvi–xvii.

38. Merchant, p. 3.

39. Institute of Medicine, National Academy of Sciences, *Women and Health Research: Ethical and Legal Issues of Including Women in Clinical Studies* (Washington, DC: National Academy Press, 1994).

40. Sue V. Rosser, Re-visioning clinical research: Gender and the ethics of experimental design, in: *Feminist Perspectives in Medical Ethics*, Helen B. Holmes and Laura Purdy, eds. (Bloomington: Indiana University Press, 1992), pp. 128–130.

41. Sharon Golub, *Periods, from Menarch to Menopause* (Newbury Park, CA: Sage Publications, 1992), p. 158.

42. Ibid., p. 162

43. African American women at extra risk, *Ms.* Jan./Feb. 1993, p. 72.

44. Gena Corea, How the new reproductive technologies will affect all women, in: *Reconstructing Babylon*, H. Patricia Hynes, ed. (Bloomington: Indiana University Press, 1991), p. 57.

Chapter 2

1. Samuel S. Peng and Jay Jaffe, Women who enter male-dominated fields of study in higher education, *American Educational Research Journal*, Vol. 16, No. 3 (Summer 1979), pp. 285–293.

2. Patricia E. White, *Women and Minorities in Science and Engineering: An Update* (Washington, DC: National Science Foundation, Jan. 1992), p. 15.

3. *How Schools Shortchange Girls* (Washington, DC: American Association of University Women Educational Foundation, 1992), p. 84.

4. Ibid., p. 86

5. Executive Summary, *Shortchanging Girls, Shortchanging America* (New York: American Association of University Women, 1994), p. 9.

6. Ibid., p. 10.

7. US Department of Education, Office for Civil Rights, *What Schools Can Do to Improve Math and Science Achievement by Minority and Female Students* (Washington, DC: US Department of Education, n.d.).

8. Bernice Taylor Anderson, How can middle schools get minority females in the math/science pipeline? *The Education Digest*, October 1993, p. 39.

9. *EQUALS Publications* (Berkeley: University of California, n.d.), p. 2.

10. For videotape ordering information, contact the Western Illinois University Curriculum Publication Clearinghouse; telephone (800) 322–3905.

11. Keith Hender, College links women in science, *Christian Science Monitor*, May 25, 1993, p. 10.

12. James Culliton, "Indirect" costs are real costs at MIT, *Science*, Feb. 14, 1992, pp. 778–779.

13. Overhead charge cap, *New York Times*, July 12, 1991, p. A17.

14. White, p. 20.

15. National Science Board, *Science and Engineering Indicators—1993* (Washington, DC: US Government Printing Office, 1993); K. Kirby and R. Czujko, The physics market: Bleak for young physicists, *Physics Today*, Vol. 46, No. 12 (Dec. 1993), pp. 22–28.

16. Linda J. Trigg and Daniel Perlman, Social influences on women's pursuit of a nontraditional career, *Psychology of Women Quarterly*, Vol. 1, No. 2 (Winter 1976), pp. 138–150.

17. Betsy Bosak Houser and Chris Garvey, The impact of family, peers, and educational personnel upon career decision making, *Journal of Vocational Behavior*, Vol. 23 (1983), pp. 35–44.

18. Mercedes S. Foster, A spouse employment program, *Bioscience*, Vol. 43 (April 1993), pp. 241–242.

19. Mary Fehrs and Roman Czujko, Women in physics: Reversing the exclusion, *Physics Today*, August 1992, pp. 33–40; Marvalee H. Wake, Two-career couples—Attitudes and opportunities, *Bioscience*, Vol. 43 (April 1993), pp. 238–240.
20. Karen Gold, Get thee to a laboratory, *New Scientist*, Vol. 14 (April 1990), p. 42.
21. Interview with Dr. Elaine Seymour, Jan. 7, 1994.
22. Boyce Rensberger, Women's place: On the podium, *Washington Post*, Aug. 26, 1992, p. A26.
23. Women in science: Discrimination against women in science is wrong but so is a quota system, opinion, *Nature*, Vol. 359 (Sept. 10, 1992), p. 92.
24. Elizabeth Kamarch Minnich, *Transforming Knowledge* (Philadelphia: Temple University Press, 1990), p. 70.
25. Constance Holden, Wanted: 675,000 future scientists and engineers, *Science*, Vol. 244, 1989, p. 761; Richard C. Atkinson, Supply and demand for scientists and engineers: A national crises in the making, *Science*, Vol. 248, 1990, pp. 425–432; Robert Pool, Who will do science in the 1990s? *Science*, Vol. 248, 1990, pp. 432–435.
26. James L. Franklin, Scientist shortage predicted in US, *Boston Globe*, Feb. 22, 1990, p. 3; Simson Garfinkel, Shortage of scientists sends signal, *Christian Science Monitor*, March 6, 1990, p. 12; Scientist shortage looms, *USA Today*, June 18, 1990.
27. P. Nulty, The hot demand for new scientists, *Fortune*, July 31, 1989, pp. 155–165.
28. Robert Crease, Bell Labs: Shakeout follows breakup, *Science*, June 14, 1991, pp. 1480–1482.
29. Gary Stix, Shrinking sandbox, *Scientific American*, Dec. 1993, pp. 45–46; David H. Freedman, A clouded future for IBM research, *Science*, April 23, 1993, pp. 480–481.
30. United States Department of Education, *The State of Education 1994* (Washington DC: National Center for Education Statistics, 1994).
31. Burton Clarke and Guy Neave, eds., *The Encyclopedia of Higher Education*, Vol. 2 (Oxford: Pergamon Press, 1992), p. 1333.
32. For a summary of the situation in physics, see: Kate Kirby and Roman Czujko, The physics job market: Bleak for young physicists, *Physics Today*, Dec. 1993, pp 22–27; for a general overview of poor employment predictions, see: Sharon Begley, Lucy Shackelford, and Adam Rogers, No PhDs need apply, *Newsweek*, Dec. 5, 1994, pp. 62–65.
33. Boyce Rensenberger, Scientist shortfall a myth, *Washington Post*, April 9, 1992, p. A1.
34. Elizabeth Fowler, Campaign aims to steer more kids toward engineering, *Star Tribune*, March 3, 1991, p. J1.
35. Lilli S. Hornig, Women in science and engineering: Why so few? *Technology Review*, Nov./Dec. 1984, reprinted in *Social Issues Resources Series*, Vol. 3, Article 19.
36. Retraining program launched to prepare former defense industry employees for careers as science and math teachers, *National Research Council News*, Dec. 21, 1994.

Chapter 3

1. Rosabeth Moss Kanter, Some effects of proportions on group life: Skewed sex ratios and responses to token women, *American Journal of Sociology*, Vol. 82, No. 5 (March 1977), pp. 965–990.
2. Stacy J. Rogers and Elizabeth G. Meneghan, Women's persistence in undergraduate majors: The effects of gender-disproportionate representation, *Gender and Society*, Vol. 5, No. 4 (December 1991), pp. 549–564.

3. Amy S. Wharton and James N. Baron, So happy together? The impact of gender segregation on men at work, *American Sociological Review*, Vol. 52 (October 1987), pp. 574–581.

4. In Minneapolis on September 22, 1994, women were encouraged to travel to the downtown pedestrian mall to greet the Wonderbra bus. The garments were so much in demand by consumers that stores' supplies were depleted and the padded-bra-laden bus's arrival was apparently cause for celebration. From an advertisement by Dayton's Department Store, The one and only Wonderbra: Special delivery today, *Star-Tribune*, Sept. 22, 1994, p. 16A.

5. *How Schools Shortchange Girls* (Washington, DC: American Association of University Women Educational Foundation, 1992).

6. *Webster's New World Dictionary*, 2nd College Ed. (Cleveland, OH: William Collins and World Publishing Company, 1976), p. 289.

7. Carolyn Mooney, Universities award record number of doctorates: More than 20% go to students from overseas, *Chronicle of Higher Education*, Vol. 37 (May 29, 1991), p. A1.

8. National Science Foundation Science Resources Studies Division, *NSF Data Brief* (Washington DC: NSF, June 20, 1994), p. 2.

9. James Marti, Darwin revisited, *Utne Reader*, March/April 1992, p. 30.

10. Jeffrey Mervis, Radcliffe president lambastes competitiveness in research, *The Scientist*, Jan. 20, 1992, p. 3.

11. Helen E. Longino and Valerie Miner, eds., A feminist taboo? in: *Competition, a Feminist Taboo?* (New York: The Feminist Press, 1987), p. 1.

12. Linda Jean Shepherd, *Lifting the Veil: The Feminine Face of Science* (Boston: Shambhala Publications, 1993), p. 207.

13. Londa Schiebinger, *The Mind Has No Sex?: Women in the Origins of Modern Science* (Cambridge, MA: Harvard University Press, 1989), p. 276.

14. Caroline Herzenberg and Ruth Howes, Women in weapons development: The Manhattan Project, in: *Women and the Use of Military Force*, Ruth H. Howes and Michael Stevenson, eds. (Boulder, CO: Lynne Rienner Publishers, 1993), pp. 95–109.

15. Ibid., p. 99.

16. Ibid. pp. 100–101.

17. A. Carol Leopold, The science community is starved for ethical standards, *The Scientist*, Jan. 6, 1992, p. 11.

18. Maria Mies, The need for a new vision: The subsistence perspective, in: *Ecofeminism* (Halifax, Canada: Fernwood Publications; London: Zed Books, 1993), pp. 302–03.

Chapter 4

1. See, for example, David F. Noble, *A World Without Women: The Christian Clerical Culture of Western Science* (New York: Alfred A. Knopf, 1992), p. 163; Margaret Alic, *Hypatia's Heritage: A History of Women in Science from Antiquity through the Nineteenth Century* (Boston: Beacon Press, 1986), pp. 62, 77; and Londa Schiebinger, *The Mind Has No Sex? Women in the Origins of Modern Science* (Cambridge, MA: Harvard University Press, 1989).

2. Pamela Abbott and Claire Wallace, *The Family and the New Right* (Boulder, CO: Pluto Press, 1992) pp. 10–11.

3. Susan Faludi, *Backlash, The Undeclared War against American Women* (New York: Crown Publishers, 1991).

4. Ibid., p. 59.

5. US Department of Commerce, Bureau of the Census, *Statistical Abstract of the United States: 1990*, 110th ed. (Washington, DC: US Government Printing Office, 1990), p. 451, Table 729.

6. Time for tots, *American Demographics*, Vol. 15, No. 8 (August 1993), p. 18.

7. Jonathan R. Cole and Harriet Zuckerman, Marriage, motherhood, and research performance, in: *The Outer Circle*, Jonathan R. Cole, Harriet Zuckerman, and John T. Bruer, eds. (New York: W.W. Norton, 1991), pp. 157–170.

8. Ibid., p. 169.

9. Amy Dacyczyn, Drowning in rising expectations, *The Tightwad Gazette*, Vol. 49 (June 1994), p. 1.

10. Michael Kinsley, Lazy he calls us, *The New Republic*, Feb. 17, 1992, p. 6.

11. Susan Chira, Custody case stirs debate on bias against working women, *New York Times*, July 31, 1994, p. 14.

12. Penelope Leach, *Children First* (New York: Alfred A. Knopf, 1994).

13. Gwen Kinkead, Spock, Brazelton and now . . . Penelope Leach, *New York Times Magazine*, April 10, 1994, p. 34.

14. Leach, p. 88.

15. Helena M. Pycior, Marie Curie's "anti-natural path": Time only for science and family, in: *Uneasy Careers and Intimate Lives: Women in Science 1789–1979*, Pnina G. Abir-Am and Dorinda Outram, eds. (New Brunswick, NJ: Rutgers University Press, 1987), p. 199.

16. Peggy K. Kidwell, Cecilia Payne-Gaposchkin: Astronomy in the family, in: *Uneasy Careers and Intimate Lives: Women in Science 1789–1979*, Pnina G. Abir-Am and Dorinda Outram, eds. (New Brunswick, NJ: Rutgers University Press, 1987), p. 232.

17. Valuing Child Care, editorial, *Christian Science Monitor*, April 22, 1994, p. 22.

18. National Science Board, Average annual salary offers to doctoral degree candidates in selected fields: 1988–93, in: *Science and Engineering Indicators—1993* (Washington, DC: US Government Printing Office, 1993), p. 77.

19. Leach, p. 98.

20. Ibid., p. 100.

21. Kinkead, p. 35.

22. Ibid.

23. Interviews with Rosemary Morse and MaryAnn Marty, March 14, 1994.

24. Open breast-feeding becomes legal right, *New York Times*, May 19, 1994, p. B4; Marylou Tousignant, Virginia passes law to protect breast-feeders, *Washington Post*, April 17, 1994, p. B5.

25. Interview with Catherine Didion, Executive Director, Association for Women in Science, July 25, 1994.

26. Rena Subotnik and Karen Arnold, Passing through the gates: Career establishment of talented women scientists, *Roeper Review*, in press.

Chapter 6

1. Helen Longino, Can there be a feminist science? in: *Feminism and Science*, Nancy Tuana, ed. (Bloomington: Indiana University Press, 1989), reprinted in *Hypatia*, Vol. 2, No. 3 (fall 1987), pp. 45–57.

2. How to pick the right candidate, *Star Tribune*, Aug. 4, 1994, p. 16A.

Chapter 8

1. Comparisons across cultures, *Science*, Vol. 263 (March 11, 1994).
2. AAUW Executive Summary, *Shortchanging Girls, Shortchanging America* (Washington, DC: American Association of University Women, 1994), p. 14.
3. Peggy Orenstein, *Schoolgirls: Young Women, Self-esteem and the Confidence Gap* (New York: Doubleday, 1994), pp. 188–189.
4. Zachary Levine, Academic community must prepare students for nontraditional careers for physicists, *APS News*, July 1994, p. 6.
5. Thomas S. Kuhn, *The Structure of Scientific Revolutions* (Chicago: The University of Chicago Press, 1970).

Index

4362047

Made in the USA
Lexington, KY
18 January 2010